# EPIOPTICS-9

## THE SCIENCE AND CULTURE SERIES — PHYSICS



1. Perspectives for New Detectors in Future Supercolliders, 1991
2. Data Structures for Particle Physics Experiments: Evolution or Revolution?, 1991
3. Image Processing for Future High-Energy Physics Detectors, 1992
4. GaAs Detectors and Electronics for High-Energy Physics, 1992
5. Supercolliders and Superdetectors, 1993
6. Properties of SUSY Particles, 1993
7. From Superstrings to Supergravity, 1994
8. Probing the Nuclear Paradigm with Heavy Ion Reactions, 1994
9. Quantum-Like Models and Coherent Effects, 1995
10. Quantum Gravity, 1996
11. Crystalline Beams and Related Issues, 1996
12. The Spin Structure of the Nucleon, 1997
13. Hadron Colliders at the Highest Energy and Luminosity, 1998
14. Universality Features in Multihadron Production and the Leading Effect, 1998
15. Exotic Nuclei, 1998
16. Spin in Gravity: Is It Possible to Give an Experimental Basis to Torsion?, 1998
17. New Detectors, 1999
18. Classical and Quantum Nonlocality, 2000
19. Silicides: Fundamentals and Applications, 2000
20. Superconducting Materials for High Energy Colliders, 2001
21. Deep Inelastic Scattering, 2001
22. Electromagnetic Probes of Fundamental Physics, 2003
23. Epioptics-7, 2004
24. Symmetries in Nuclear Structure, 2004
25. Innovative Detectors for Supercolliders, 2003
26. Complexity, Metastability and Nonextensivity, 2004
27. Epioptics-8, 2004
28. The Physics and Applications of High Brightness Electron Beams, 2005
29. Epioptics-9, 2006

THE SCIENCE AND CULTURE SERIES – PHYSICS

# EPIOPTICS-9

Proceedings of the 39th Course of the
International School of Solid State Physics

Erice, Italy 20 — 26 July 2006

Editor
**Antonio Cricenti**

Series Editor
**A. Zichichi**

World Scientific

NEW JERSEY • LONDON • SINGAPORE • BEIJING • SHANGHAI • HONG KONG • TAIPEI • CHENNAI

*Published by*

World Scientific Publishing Co. Pte. Ltd.
5 Toh Tuck Link, Singapore 596224
*USA office:* 27 Warren Street, Suite 401-402, Hackensack, NJ 07601
*UK office:* 57 Shelton Street, Covent Garden, London WC2H 9HE

**British Library Cataloguing-in-Publication Data**
A catalogue record for this book is available from the British Library.

**The Science and Culture Series — Physics**
**EPIOPTICS-9**
**Proceedings of the 39th Course of the International School of Solid State Physics**


ISBN-13 978-981-279-402-4
ISBN-10 981-279-402-6

Printed in Singapore by B & JO Enterprise

# PREFACE

This special volume contains the Proceedings of the 9th Epioptics Workshop, held in the Ettore Majorana Foundation and Centre for Scientific Culture, Erice, Sicily, from July 20 to 26, 2006. The Workshop was the 9th in the Epioptics series and the 39th of the International School of Solid State Physics. Antonio Cricenti from CNR Istituto di Struttura della Materia and Theo Rasing from the University of Njimegen, were the Directors of the Workshop. The Advisory Committee of the Workshop included Y. Borensztein from U. Paris VII (F), R. Del Sole from U. Roma II Tor Vergata (I), D. Aspnes from NCSU (USA), O. Hunderi from U. Trondheim (N), J. McGilp from Trinity College Dublin (Eire), W. Richter from TU Berlin (D), N. Tolk from Vanderbilt University (USA), and P. Weightman from University Liverpool (UK). Fifty five scientists from sixteen countries attended the Workshop.

The Workshop has brought together researchers from universities and research institutes who work in the fields of (semiconductor) surface science, epitaxial growth, materials deposition and optical diagnostics relevant to (semiconductor) materials and structures of interest for present and anticipated (spin) electronic devices. The Workshop was aimed at assessing the capabilities of state-of-the-art optical techniques in elucidating the fundamental electronic and structural properties of semiconductor and metal surfaces, interfaces, thin layers, and layer structures, and assessing the usefulness of these techniques for optimization of high quality multilayer samples through feedback control during materials growth and processing. Particular emphasis is dedicated to theory of non-linear optics and to dynamical processes through the use of pump-probe techniques together with the search for new optical sources. Some new applications of Scanning Probe Microscopy to material science and biological samples, dried and *in vivo*, with the use of different laser sources have also been presented. Materials of particular interest have been silicon, semiconductor-metal interfaces, semiconductor and magnetic multi-layers and III-V compound semiconductors. The Workshop is characterized by the adequate collection of notes in this volume, combined with the tutorials in some of the most advanced topics in the field.

This book is dedicated to Professor Gianfranco Chiarotti for his fundamental contributions to the development of Optical Spectroscopy as a tool to study Surface States: these studies have paved the way to the establishment of our Epioptics Community. During the School Prof. Chiarotti has been awarded the diploma of Father of Epioptics School, for the lecture regarding his contribution to the discovery of "Optical transition between semiconductor surface states". Prof. Giorgio Benedek was also awarded in the occasion of his 65th birthday for his strong and successful effort in running the International School of Solid State Physics.

I want personally to thank Prof. Chiarotti for giving me the opportunity to start this wonderful trip in science and as an example of rigour and dedication in the endeavour of scientific research. It is sad to remember that during the editing of these Proceedings two of our colleagues passed away: Dr. Marco Fabio Righini, who started his research at CNR in our optical group becoming an excellent scientist and with whom I shared the optical set-up for several years, and Prof. Carlo Coluzza, who was a very intuitive scientist in many different fields of science and a very good friend. I want also to remember my father-in-law Benito Cello, who had been very special to me, and my father Domenico, who also passed away recently: he was an untiring worker all his life, who has been exemplary and has been a great resource for me.

I wish to thank our sponsors, the Italian National Research Council (CNR) and the Sicilian Regional Government for facilitating a most successful Workshop. We wish to thank Prof. A. Zichichi, the President of the Ettore Majorana Foundation and Director of the Ettore Majorana Centre for Scientific Culture in Erice, and all the staff members of the Centre for the excellent support, organization and hospitality provided.

*Antonio Cricenti*

# CONTENTS

# LONGITUDINAL GAUGE THEORY OF SURFACE SECOND HARMONIC GENERATION

BERNARDO S. MENDOZA
*Centro de Investigaciones en Optica*
*León, Guanajuato, México*
*bms@cio.mx*

A theoretical review of surface second harmonic generation from semiconductor surfaces based on the longitudinal gauge is presented. The so called, layer-by-layer analysis is carefully presented in order to show how a surface calculation of second harmonic generation (SHG) can readily be carried out. The nonlinear susceptibility tensor $\boldsymbol{\chi}$ is split into two terms, one that is related to inter-band one-electron transitions, and the other is related to intra-band one-electron transitions. The equivalence of this formulation to the transverse gauge approach is shown and the possibility of confirming its numerical accuracy is discussed. Also, the calculation of the surface second harmonic radiated intensity $R$ within the three-layer-model is derived. With $\boldsymbol{\chi}$ and $R$ one has a complete description of this fascinating optical phenomena.

## 1. Introduction

Second harmonic generation (SHG) has become a powerful spectroscopic tool to study optical properties of surfaces and interfaces since it has the advantage of being surface sensitive. For centrosymmetric materials inversion symmetry forbids, within the dipole approximation, SHG from the bulk, but it is allowed at the surface, where the inversion symmetry is broken. Therefore, SHG should necessarily come from a localized surface region. SHG allows to study the structural atomic arrangement and phase transitions of clean and adsorbate covered surfaces, and since it is an optical probe, it can be used out of UHV conditions, and is non-invasive and non-destructive. On the experimental side, the new tunable high intensity laser systems have made SHG spectroscopy readily accessible and applicable to a wide range of systems.[1] However, the theoretical development of the field is still an ongoing subject of research. Some recent advances for the case of semiconducting and metallic systems have appeared in the literature, where the confrontation of theoretical models with experiment has yield correct physical interpretations for the SHG spectra. [1,2,3,4,5,6,7,8]

In a previous article,[9] we reviewed some of the recent results in the study of SHG using the transverse gauge for the coupling between the electromagnetic field and the electron. In particular, we showed a method to systematically investigate the different contributions to the observed peaks in SHG.[10] The approach consisted in the separation of the different contributions to the nonlinear susceptibility according to $1\omega$ and $2\omega$ transitions and to the surface or bulk character of the states among which the transitions take place. To complement above results, on this article we review the calculation of the nonlinear susceptibility using the longitudinal gauge, and show that both gauges give, as they should, the same result. We discuss a possible numerical check up on this equivalency. Also, the so called three-layer-model for the calculation of the surface radiated SH efficiency is presented.

## 2. Longitudinal Gauge

To calculate the optical properties of a given system within the longitudinal gauge, we follow the article by Aversa and Sipe.[11] A more recent derivation can also be found in Ref. [12] and [13]. Assuming the long-wavelength approximation, which implies a position independent electric field, the hamiltonian in the so called length gauge approximation is given by

$$\hat{H} = \hat{H}_0 - e\hat{\mathbf{r}} \cdot \mathbf{E}, \tag{1}$$

where $H_0 = p^2/2m + V(\mathbf{r})$, where $V(\mathbf{r}) = V(\mathbf{r} + \mathbf{R})$ is the periodic crystal potential, with $\mathbf{R}$ the real-space lattice vector. The electric field $\mathbf{E} = -\dot{\mathbf{A}}/c$, with $\mathbf{A}$ the vector potential. $H_0$ has eigenvalues $\hbar\omega_n(\mathbf{k})$ and eigenvectors $|n\mathbf{k}\rangle$ (Bloch states) labeled by a band index $n$ and crystal momentum $\mathbf{k}$. The $r$ representation of the Bloch states is given by

$$\psi_{n\mathbf{k}}(\mathbf{r}) = \langle \mathbf{r}|n\mathbf{k}\rangle = \sqrt{\frac{\Omega}{8\pi^3}} e^{i\mathbf{k}\cdot\mathbf{r}} u_{n\mathbf{k}}(\mathbf{r}), \tag{2}$$

where $u_{n\mathbf{k}}(\mathbf{r}) = u_{n\mathbf{k}}(\mathbf{r} + \mathbf{R})$ is cell periodic, and

$$\int_\Omega d^3r\, u^*_{n\mathbf{k}}(\mathbf{r}) u_{m\mathbf{q}}(\mathbf{r}) = \delta_{nm}\delta_{\mathbf{k},\mathbf{q}}, \tag{3}$$

with $\Omega$ the volume of the unit cell.

The key ingredient in the calculation are the matrix elements of the position operator $\mathbf{r}$, so we start from the basic relation

$$\langle n\mathbf{k}|m\mathbf{k}'\rangle = \delta_{nm}\delta(\mathbf{k} - \mathbf{k}'), \tag{4}$$

and take its derivative with respect to $\mathbf{k}$ as follows. On one hand,

$$\frac{\partial}{\partial \mathbf{k}}\langle n\mathbf{k}|m\mathbf{k}'\rangle = \delta_{nm}\frac{\partial}{\partial \mathbf{k}}\delta(\mathbf{k}-\mathbf{k}'), \tag{5}$$

on the other,

$$\begin{aligned}\frac{\partial}{\partial \mathbf{k}}\langle n\mathbf{k}|m\mathbf{k}'\rangle &= \frac{\partial}{\partial \mathbf{k}}\int d\mathbf{r}\langle n\mathbf{k}|\mathbf{r}\rangle\langle\mathbf{r}|m\mathbf{k}'\rangle \\ &= \int d\mathbf{r}\left(\frac{\partial}{\partial \mathbf{k}}\psi^*_{n\mathbf{k}}(\mathbf{r})\right)\psi_{m\mathbf{k}'}(\mathbf{r}),\end{aligned} \tag{6}$$

the derivative of the wavefunction is simply given by

$$\frac{\partial}{\partial \mathbf{k}}\psi^*_{n\mathbf{k}}(\mathbf{r}) = \sqrt{\frac{\Omega}{8\pi^3}}\left(\frac{\partial}{\partial \mathbf{k}}u^*_{n\mathbf{k}}(\mathbf{r})\right)e^{-i\mathbf{k}\cdot\mathbf{r}} - i\mathbf{r}\psi^*_{n\mathbf{k}}(\mathbf{r}). \tag{7}$$

We take this back into Eq. 6, to obtain

$$\begin{aligned}\frac{\partial}{\partial \mathbf{k}}\langle n\mathbf{k}|m\mathbf{k}'\rangle &= \sqrt{\frac{\Omega}{8\pi^3}}\int d\mathbf{r}\left(\frac{\partial}{\partial \mathbf{k}}u^*_{n\mathbf{k}}(\mathbf{r})\right)e^{-i\mathbf{k}\cdot\mathbf{r}}\psi_{m\mathbf{k}'}(\mathbf{r}) \\ &\quad -i\int d\mathbf{r}\psi^*_{n\mathbf{k}}(\mathbf{r})\mathbf{r}\psi_{m\mathbf{k}'}(\mathbf{r}) \\ &= \frac{\Omega}{8\pi^3}\int d\mathbf{r}\, e^{-i(\mathbf{k}-\mathbf{k}')\cdot\mathbf{r}}\left(\frac{\partial}{\partial \mathbf{k}}u^*_{n\mathbf{k}}(\mathbf{r})\right)u_{m\mathbf{k}'}(\mathbf{r}) \\ &\quad -i\langle n\mathbf{k}|\hat{\mathbf{r}}|m\mathbf{k}'\rangle.\end{aligned} \tag{8}$$

Restricting $\mathbf{k}$ and $\mathbf{k}'$ to the first Brillouin zone, we use the following valid result for any periodic function $f(\mathbf{r}) = f(\mathbf{r}+\mathbf{R})$ (see Appendix A),

$$\int d^3r\, e^{i(\mathbf{q}-\mathbf{k})\cdot\mathbf{r}}f(\mathbf{r}) = \frac{8\pi^3}{\Omega}\delta(\mathbf{q}-\mathbf{k})\int_\Omega d^3r\, f(\mathbf{r}), \tag{9}$$

to finally write,[14]

$$\begin{aligned}\frac{\partial}{\partial \mathbf{k}}\langle n\mathbf{k}|m\mathbf{k}'\rangle &= \delta(\mathbf{k}-\mathbf{k}')\int_\Omega d\mathbf{r}\left(\frac{\partial}{\partial \mathbf{k}}u^*_{n\mathbf{k}}(\mathbf{r})\right)u_{m\mathbf{k}}(\mathbf{r}) \\ &\quad -i\langle n\mathbf{k}|\hat{\mathbf{r}}|m\mathbf{k}'\rangle.\end{aligned} \tag{10}$$

where $\Omega$ is the volume of the unit cell. From

$$\int_\Omega u_{m\mathbf{k}}u^*_{n\mathbf{k}}d\mathbf{r} = \delta_{nm}, \tag{11}$$

we easily find that

$$\int_\Omega d\mathbf{r}\left(\frac{\partial}{\partial \mathbf{k}}u_{m\mathbf{k}}(\mathbf{r})\right)u^*_{n\mathbf{k}}(\mathbf{r}) = -\int_\Omega d\mathbf{r}\, u_{m\mathbf{k}}(\mathbf{r})\left(\frac{\partial}{\partial \mathbf{k}}u^*_{n\mathbf{k}}(\mathbf{r})\right). \tag{12}$$

Therefore, we define

$$\boldsymbol{\xi}_{nm}(\mathbf{k}) \equiv i \int_{\Omega} d\mathbf{r}\, u^*_{n\mathbf{k}}(\mathbf{r}) \nabla_{\mathbf{k}} u_{m\mathbf{k}}(\mathbf{r}), \tag{13}$$

with $\partial/\partial\mathbf{k} = \nabla_{\mathbf{k}}$. Now, from Eqs. 5, 8, and 13, we have that the matrix elements of the position operator of the electron are given by

$$\langle n\mathbf{k}|\hat{\mathbf{r}}|m\mathbf{k}'\rangle = \delta(\mathbf{k}-\mathbf{k}')\boldsymbol{\xi}_{nm}(\mathbf{k}) + i\delta_{nm}\nabla_{\mathbf{k}}\delta(\mathbf{k}-\mathbf{k}'), \tag{14}$$

Then, from Eq. (14), and writing $\hat{\mathbf{r}} = \hat{\mathbf{r}}_e + \hat{\mathbf{r}}_i$, with $\hat{\mathbf{r}}_e$ ($\hat{\mathbf{r}}_i$) the interband (intraband) part, we obtain that

$$\langle n\mathbf{k}|\hat{\mathbf{r}}_i|m\mathbf{k}'\rangle = \delta_{nm}\left[\delta(\mathbf{k}-\mathbf{k}')\boldsymbol{\xi}_{nn}(\mathbf{k}) + i\nabla_{\mathbf{k}}\delta(\mathbf{k}-\mathbf{k}')\right], \tag{15}$$

$$\langle n\mathbf{k}|\hat{\mathbf{r}}_e|m\mathbf{k}'\rangle = (1-\delta_{nm})\delta(\mathbf{k}-\mathbf{k}')\boldsymbol{\xi}_{nm}(\mathbf{k}). \tag{16}$$

To proceed, we relate Eq. 16 to the matrix elements of the momentum operator as follows. We start from the basic relation,

$$\hat{\mathbf{v}} = \frac{1}{i\hbar}[\hat{\mathbf{r}}, \hat{H}_0], \tag{17}$$

with $\hat{\mathbf{v}}$ the velocity operator. Neglecting nonlocal potentials in $\hat{H}_0$ we obtain, on one hand

$$[\hat{\mathbf{r}}, \hat{H}_0] = i\hbar\frac{\hat{\mathbf{p}}}{m}, \tag{18}$$

with $\hat{\mathbf{p}}$ the momentum operator, with $m$ the mass of the electron. On the other hand,

$$\langle n\mathbf{k}|[\hat{\mathbf{r}}, \hat{H}_0]|m\mathbf{k}\rangle = \langle n\mathbf{k}|\hat{\mathbf{r}}\hat{H}_0 - \hat{H}_0\hat{\mathbf{r}}|m\mathbf{k}\rangle = (\hbar\omega_m(\mathbf{k}) - \hbar\omega_n(\mathbf{k}))\langle n\mathbf{k}|\hat{\mathbf{r}}|m\mathbf{k}\rangle, \tag{19}$$

thus defining $\omega_{nm\mathbf{k}} = \omega_n(\mathbf{k}) - \omega_m(\mathbf{k})$ we get

$$\mathbf{r}_{nm}(\mathbf{k}) = \frac{\mathbf{p}_{nm}(\mathbf{k})}{im\omega_{nm}(\mathbf{k})} = \frac{\mathbf{v}_{nm}(\mathbf{k})}{i\omega_{nm}(\mathbf{k})} \qquad n \neq m. \tag{20}$$

Comparing above result with Eq. 16, we can identify

$$(1-\delta_{nm})\boldsymbol{\xi}_{nm} \equiv \mathbf{r}_{nm}, \tag{21}$$

and the we can write

$$\langle n\mathbf{k}|\hat{\mathbf{r}}_e|m\mathbf{k}\rangle = \mathbf{r}_{nm}(\mathbf{k}) = \frac{\mathbf{p}_{nm}(\mathbf{k})}{im\omega_{nm}(\mathbf{k})} \qquad n \neq m, \tag{22}$$

which gives the interband matrix elements of the position operator in terms of the matrix elements of the well defined momentum operator.

For the intraband part, we derive the following general result,

$$
\begin{aligned}
\langle n\mathbf{k}|[\hat{\mathbf{r}}_i,\hat{\mathcal{O}}]|m\mathbf{k}'\rangle &= \sum_{\ell,\mathbf{k}''}\Big(\langle n\mathbf{k}|\hat{\mathbf{r}}_i|\ell\mathbf{k}''\rangle\langle \ell\mathbf{k}''|\hat{\mathcal{O}}|m\mathbf{k}'\rangle \\
&\quad -\langle n\mathbf{k}|\hat{\mathcal{O}}|\ell\mathbf{k}''\rangle\langle \ell\mathbf{k}''|\hat{\mathbf{r}}_i|m\mathbf{k}'\rangle\Big) \\
&= \sum_{\ell}\left(\langle n\mathbf{k}|\hat{\mathbf{r}}_i|\ell\mathbf{k}'\rangle\mathcal{O}_{\ell m}(\mathbf{k}')\right. \\
&\quad \left.-\mathcal{O}_{n\ell}(\mathbf{k})|\ell\mathbf{k}\rangle\langle \ell\mathbf{k}|\hat{\mathbf{r}}_i|m\mathbf{k}'\rangle\right),
\end{aligned} \tag{23}
$$

where we have taken $\langle n\mathbf{k}|\hat{\mathcal{O}}|\ell\mathbf{k}''\rangle = \delta(\mathbf{k}-\mathbf{k}'')\mathcal{O}_{n\ell}(\mathbf{k})$. We substitute Eq. 15, to obtain

$$
\begin{aligned}
\sum_{\ell} \quad & \left(\delta_{n\ell}\left[\delta(\mathbf{k}-\mathbf{k}')\boldsymbol{\xi}_{nn}(\mathbf{k})+i\nabla_{\mathbf{k}}\delta(\mathbf{k}-\mathbf{k}')\right]\mathcal{O}_{\ell m}(\mathbf{k}')\right. \\
& \left.-\mathcal{O}_{n\ell}(\mathbf{k})\delta_{\ell m}\left[\delta(\mathbf{k}-\mathbf{k}')\boldsymbol{\xi}_{mm}(\mathbf{k})+i\nabla_{\mathbf{k}}\delta(\mathbf{k}-\mathbf{k}')\right]\right) \\
&= \left(\left[\delta(\mathbf{k}-\mathbf{k}')\boldsymbol{\xi}_{nn}(\mathbf{k})+i\nabla_{\mathbf{k}}\delta(\mathbf{k}-\mathbf{k}')\right]\mathcal{O}_{nm}(\mathbf{k}')\right. \\
&\quad \left.-\ \mathcal{O}_{nm}(\mathbf{k})\left[\delta(\mathbf{k}-\mathbf{k}')\boldsymbol{\xi}_{mm}(\mathbf{k})+i\nabla_{\mathbf{k}}\delta(\mathbf{k}-\mathbf{k}')\right]\right) \\
&= \delta(\mathbf{k}-\mathbf{k}')\mathcal{O}_{nm}(\mathbf{k})\left(\boldsymbol{\xi}_{nn}(\mathbf{k})-\boldsymbol{\xi}_{mm}(\mathbf{k})\right)+i\mathcal{O}_{nm}(\mathbf{k}')\nabla_{\mathbf{k}}\delta(\mathbf{k}-\mathbf{k}') \\
&\quad +i\delta(\mathbf{k}-\mathbf{k}')\nabla_{\mathbf{k}}\mathcal{O}_{nm}(\mathbf{k})-i\mathcal{O}_{nm}(\mathbf{k}')\nabla_{\mathbf{k}}\delta(\mathbf{k}-\mathbf{k}') \\
&= i\delta(\mathbf{k}-\mathbf{k}')\Big(\nabla_{\mathbf{k}}\mathcal{O}_{nm}(\mathbf{k})-i\mathcal{O}_{nm}(\mathbf{k})\left(\boldsymbol{\xi}_{nn}(\mathbf{k})-\boldsymbol{\xi}_{mm}(\mathbf{k})\right)\Big) \\
&\equiv i\delta(\mathbf{k}-\mathbf{k}')(\mathcal{O}_{nm})_{;\mathbf{k}}.
\end{aligned} \tag{24}
$$

Then,

$$
\langle n\mathbf{k}|[\hat{\mathbf{r}}_i,\hat{\mathcal{O}}]|m\mathbf{k}'\rangle = i\delta(\mathbf{k}-\mathbf{k}')(\mathcal{O}_{nm})_{;\mathbf{k}}, \tag{25}
$$

with

$$
(\mathcal{O}_{nm})_{;\mathbf{k}} = \nabla_{\mathbf{k}}\mathcal{O}_{nm}(\mathbf{k})-i\mathcal{O}_{nm}(\mathbf{k})\left(\boldsymbol{\xi}_{nn}(\mathbf{k})-\boldsymbol{\xi}_{mm}(\mathbf{k})\right), \tag{26}
$$

the generalized derivative of $\mathcal{O}_{nm}$ with respect to $\mathbf{k}$. Note that the highly singular term $\nabla_{\mathbf{k}}\delta(\mathbf{k}-\mathbf{k}')$ cancels in Eq. 24, thus giving a well defined commutator of the intraband position operator with an arbitrary operator $\hat{\mathcal{O}}$. We use Eq. 22 and 25 in the next section.

## 3. Time-dependent Perturbation Theory

We use, in the independent particle approximation, the electron density operator $\hat{\rho}$ to obtain, the expectation value of any observable $\mathcal{O}$ as

$$
\mathcal{O} = Tr(\hat{\mathcal{O}}\hat{\rho}) = Tr(\hat{\rho}\hat{\mathcal{O}}), \tag{27}
$$

where $Tr$ is the trace, that as we have shown has the property of being invariant under cyclic permutations. The dynamical equation of motion for $\rho$ is given by

$$i\hbar\frac{d\hat{\rho}}{dt} = [\hat{H}, \hat{\rho}], \tag{28}$$

where it is more convenient to work in the interaction picture, for which we transform all the operators according to

$$\hat{\mathcal{O}}_I = \hat{U}\hat{\mathcal{O}}\hat{U}^\dagger, \tag{29}$$

where

$$\hat{U} = e^{i\hat{H}_0 t/\hbar}, \tag{30}$$

is the unitary operator that take us to the interaction picture. Note that $\hat{\mathcal{O}}_I$ depends on time even if $\hat{\mathcal{O}}$ does not. Then, we transform Eq. 28 into

$$i\hbar\frac{d\hat{\rho}_I(t)}{dt} = [-e\hat{\mathbf{r}}_I(t)\cdot\mathbf{E}(t), \hat{\rho}_I(t)], \tag{31}$$

that leads to

$$\hat{\rho}_I(t) = \hat{\rho}_I(t=-\infty) + \frac{ie}{\hbar}\int_{-\infty}^{t} dt'[\hat{\mathbf{r}}_I(t')\cdot\mathbf{E}(t'), \hat{\rho}_I(t')]. \tag{32}$$

We assume that the interaction is switched-on adiabatically, and choose a time-periodic perturbing field, to write

$$\mathbf{E}(t) = \mathbf{E}e^{-i\omega t}e^{\eta t}, \tag{33}$$

where $\eta > 0$ assures that at $t = -\infty$ the interaction is zero and has its full strength, $\mathbf{E}$, at $t = 0$. After the required time integrals are done, one takes $\eta \to 0$. Instead of Eq. 33 we use

$$\mathbf{E}(t) = \mathbf{E}e^{-i\tilde{\omega}t}, \tag{34}$$

with

$$\tilde{\omega} = \omega + i\eta. \tag{35}$$

Also, $\hat{\rho}_I(t=-\infty)$ should be independent of time, and thus $[\hat{H}, \hat{\rho}]_{t=-\infty} = 0$, which implies that $\hat{\rho}_I(t=-\infty) = \hat{\rho}(t=-\infty) \equiv \hat{\rho}_0$, where $\hat{\rho}_0$ is the density matrix of the unperturbed ground state, such that

$$\langle n\mathbf{k}|\hat{\rho}_0|m\mathbf{k}'\rangle = f_n(\hbar\omega_n(\mathbf{k}))\delta_{nm}\delta(\mathbf{k}-\mathbf{k}'), \tag{36}$$

where $f_n(\hbar\omega_n(\mathbf{k})) = f_{n\mathbf{k}}$ is the Fermi-Dirac distribution function.

We solve Eq. 32 using the standard iterative solution, for which we write

$$\hat{\rho}_I = \hat{\rho}_I^{(0)} + \hat{\rho}_I^{(1)} + \hat{\rho}_I^{(2)} + \cdots, \tag{37}$$

where $\hat{\rho}_I^{(N)}$ is the density operator to order $N$ in $\mathbf{E}(t)$. Then, Eq. 32 reads

$$\hat{\rho}_I^{(0)}+\hat{\rho}_I^{(1)}+\hat{\rho}_I^{(2)}+\cdots=\hat{\rho}_0+\frac{ie}{\hbar}\int_{-\infty}^{t}dt'[\hat{\mathbf{r}}_I(t')\cdot\mathbf{E}(t'),\hat{\rho}_I^{(0)}+\hat{\rho}_I^{(1)}+\hat{\rho}_I^{(2)}+\cdots],\tag{38}$$

where by equating equal orders in the perturbation, we find

$$\hat{\rho}_I^{(0)}\equiv\hat{\rho}_0,\tag{39}$$

and

$$\hat{\rho}_I^{(N)}(t)=\frac{ie}{\hbar}\int_{-\infty}^{t}dt'[\hat{\mathbf{r}}_I(t')\cdot\mathbf{E}(t'),\hat{\rho}_I^{(N-1)}(t')].\tag{40}$$

It is simple to show that matrix elements of Eq. (40) satisfy $\langle n\mathbf{k}|\rho_I^{(N+1)}(t)|m\mathbf{k}'\rangle=\rho_{I,nm}^{(N+1)}(\mathbf{k})\delta(\mathbf{k}-\mathbf{k}')$, with

$$\rho_{I,nm}^{(N+1)}(\mathbf{k};t)=\frac{ie}{\hbar}\int_{-\infty}^{t}dt'\langle n\mathbf{k}|[\hat{\mathbf{r}}_I(t'),\hat{\rho}_I^{(N)}(t')]|m\mathbf{k}\rangle\cdot\mathbf{E}(t').\tag{41}$$

Now we work out the commutator of Eq. 41. Then,

$$\begin{aligned}\langle n\mathbf{k}|[\hat{\mathbf{r}}_I(t),\hat{\rho}_I^{(N)}(t)]|m\mathbf{k}\rangle&=\langle n\mathbf{k}|[\hat{U}\hat{\mathbf{r}}\hat{U}^\dagger,\hat{U}\hat{\rho}^{(N)}(t)\hat{U}^\dagger]|m\mathbf{k}\rangle\\&=\langle n\mathbf{k}|\hat{U}[\hat{\mathbf{r}},\hat{\rho}^{(N)}(t)]\hat{U}^\dagger|m\mathbf{k}\rangle\\&=e^{i\omega_{nm\mathbf{k}}t}\left(\langle n\mathbf{k}|[\hat{\mathbf{r}}_e,\hat{\rho}^{(N)}(t)]+[\hat{\mathbf{r}}_i,\hat{\rho}^{(N)}(t)]|m\mathbf{k}\rangle\right),\end{aligned}\tag{42}$$

where the time dependence of operator's interaction picture is explicitly shown by the exponential factor, and the implicit dependence of $\hat{\rho}^{(N)}$ inherited from Eq. 28 is shown by its $t$ argument. We calculate the interband term first, so using Eq. 22 we obtain

$$\begin{aligned}\langle n\mathbf{k}|[\hat{\mathbf{r}}_e,\hat{\rho}^{(N)}(t)]|m\mathbf{k}\rangle&=\sum_{\ell}\Big(\langle n\mathbf{k}|\hat{\mathbf{r}}_e|\ell\mathbf{k}\rangle\langle\ell\mathbf{k}|\hat{\rho}^{(N)}(t)|m\mathbf{k}\rangle\\&\quad-\langle n\mathbf{k}|\hat{\rho}^{(N)}(t)|\ell\mathbf{k}\rangle\langle\ell\mathbf{k}|\hat{\mathbf{r}}_e|m\mathbf{k}\rangle\Big)\\&=\sum_{\ell\neq n,m}\Big(\mathbf{r}_{n\ell}(\mathbf{k})\rho_{\ell m}^{(N)}(\mathbf{k};t)-\rho_{n\ell}^{(N)}(\mathbf{k};t)\mathbf{r}_{\ell m}(\mathbf{k})\Big)\\&\equiv\mathbf{R}_e^{(N)}(\mathbf{k};t).\end{aligned}\tag{43}$$

Now, from Eq. 25 we simply obtain,

$$\langle n\mathbf{k}|[\hat{\mathbf{r}}_i,\hat{\rho}^{(N)}(t)]|m\mathbf{k}'\rangle=i(\rho_{nm}^{(N)}(t))_{;\mathbf{k}}\equiv\mathbf{R}_i^{(N)}(\mathbf{k};t).\tag{44}$$

Then Eq. 41 becomes,

$$\rho_{I,nm}^{(N+1)}(\mathbf{k};t)=\frac{ie}{\hbar}\int_{-\infty}^{t}dt'e^{i(\omega_{nm\mathbf{k}}-\tilde{\omega})t'}\left[R_e^{b(N)}(\mathbf{k};t')+R_i^{b(N)}(\mathbf{k};t')\right]E^b,\tag{45}$$

where, the roman superindices $a, b, c$ denote Cartesian components that are summed over if repeated. We start with the linear response, then from Eq. 36 and 43,[a]

$$R_e^{b(0)}(\mathbf{k};t) = \sum_\ell \left( r^b_{n\ell}(\mathbf{k})\rho^{(0)}_{\ell m}(\mathbf{k}) - \rho^{(0)}_{n\ell}(\mathbf{k}) r^b_{\ell m}(\mathbf{k}) \right)$$
$$= \sum_\ell \left( r^b_{n\ell}(\mathbf{k})\delta_{\ell m} f_m(\hbar\omega_m(\mathbf{k})) - \delta_{n\ell} f_n(\hbar\omega_n(\mathbf{k})) r^b_{\ell m}(\mathbf{k}) \right)$$
$$= f_{mn\mathbf{k}} r^b_{nm}(\mathbf{k}), \tag{46}$$

where $f_{mn\mathbf{k}} = f_{m\mathbf{k}} - f_{n\mathbf{k}}$. Also, from Eq. 44 and Eq. 26

$$R_i^{b(0)}(\mathbf{k}) = i(\rho^{(0)}_{nm})_{;k^b} = i\delta_{nm}(f_{n\mathbf{k}})_{;k^b} = i\delta_{nm}\nabla_{k^b} f_{n\mathbf{k}}. \tag{47}$$

For a semiconductor at $T = 0$, $f_{n\mathbf{k}}$ is one if the state $|n\mathbf{k}\rangle$ is a valence state and zero if it is a conduction state, thus $\nabla_{\mathbf{k}} f_{n\mathbf{k}} = 0$ and $\mathbf{R}_i^{(0)} = 0$. Therefore the linear response has no contribution from intraband transitions. Then,

$$\rho^{(1)}_{I,nm}(\mathbf{k};t) = \frac{ie}{\hbar} f_{mn\mathbf{k}} r^b_{nm}(\mathbf{k}) E^b \int_{-\infty}^t dt' e^{i(\omega_{nm\mathbf{k}} - \tilde{\omega})t'}$$
$$= \frac{e}{\hbar} f_{mn\mathbf{k}} r^b_{nm}(\mathbf{k}) E^b \frac{e^{i(\omega_{nm\mathbf{k}} - \tilde{\omega})t}}{\omega_{nm\mathbf{k}} - \tilde{\omega}}$$
$$= e^{i\omega_{nm\mathbf{k}}t} B^b_{mn}(\mathbf{k}) E^b(t)$$
$$= e^{i\omega_{nm\mathbf{k}}t} \rho^{(1)}_{nm}(\mathbf{k};t). \tag{48}$$

We generalize this result since we need it for the non-linear response. In general we could have several perturbing fields with different frequencies, i.e. $\mathbf{E}(t) = \mathbf{E}_{\omega_\alpha} e^{-i\tilde{\omega}_\alpha t}$, then

$$\rho^{(1)}_{nm}(\mathbf{k};t) = B^b_{mn}(\mathbf{k},\omega_\alpha) E^b_{\omega_\alpha} e^{-i\tilde{\omega}_\alpha t}, \tag{49}$$

with

$$B^b_{nm}(\mathbf{k},\omega_\alpha) = \frac{e}{\hbar} \frac{f_{mn\mathbf{k}} r^b_{nm}(\mathbf{k})}{\omega_{nm\mathbf{k}} - \tilde{\omega}_\alpha}. \tag{50}$$

Now, we calculate the second-order response. Then, from Eq. 43

$$R_e^{b(1)}(\mathbf{k};t) = \sum_\ell \left( r^b_{n\ell}(\mathbf{k})\rho^{(1)}_{\ell m}(\mathbf{k};t) - \rho^{(1)}_{n\ell}(\mathbf{k};t) r^b_{\ell m}(\mathbf{k}) \right)$$
$$= \sum_\ell \left( r^b_{n\ell}(\mathbf{k}) B^c_{\ell m}(\mathbf{k},\omega_\beta) - B^c_{n\ell}(\mathbf{k},\omega_\beta) r^b_{\ell m}(\mathbf{k}) \right) E^c_{\omega_\beta}(t), \tag{51}$$

and from Eq. 44

$$R_i^{b(1)}(\mathbf{k};t) = i(\rho^{(1)}_{nm}(t))_{;k^b} = iE^c_{\omega_\beta}(t)(B^c_{nm}(\mathbf{k},\omega_\beta))_{;k^b}. \tag{52}$$

[a] from now on, it should be clear that the matrix elements of $\mathbf{r}_{nm}$ imply $n \neq m$.

Using Eqs. 51 and 52 in Eq. (45), and generalizing to two different perturbing fields, we obtain

$$\begin{aligned}\rho^{(2)}_{I,nm}(\mathbf{k};t) &= \frac{ie}{\hbar}\Big[\sum_{\ell}\Big(r^b_{n\ell}(\mathbf{k})B^c_{\ell m}(\mathbf{k},\omega_\beta) - B^c_{n\ell}(\mathbf{k},\omega_\beta)r^b_{\ell m}(\mathbf{k})\Big) \\ &\quad + i(B^c_{nm}(\mathbf{k},\omega_\beta))_{;k^b}\Big]E^b_{\omega_\alpha}E^c_{\omega_\beta}\int_{-\infty}^{t} dt' e^{i(\omega_{nm\mathbf{k}}-\tilde{\omega}_\alpha-\tilde{\omega}_\beta)t'} \\ &= \frac{e}{\hbar}\Big[\sum_{\ell}\Big(r^b_{n\ell}(\mathbf{k})B^c_{\ell m}(\mathbf{k},\omega_\beta) - B^c_{n\ell}(\mathbf{k},\omega_\beta)r^b_{\ell m}(\mathbf{k})\Big) \\ &\quad + i(B^c_{nm}(\mathbf{k},\omega_\beta))_{;k^b}\Big]E^b_{\omega_\alpha}E^c_{\omega_\beta}\frac{e^{i(\omega_{nm\mathbf{k}}-\tilde{\omega}_3)t}}{\omega_{nm\mathbf{k}}-\tilde{\omega}_3} \\ &= e^{i\omega_{nm\mathbf{k}}t}\rho^{(2)}_{nm}(\mathbf{k};t).\end{aligned} \tag{53}$$

Now, we write $\rho^{(2)}_{nm}(\mathbf{k};t) = \rho^{(2)}_{nm}(\mathbf{k};\omega_3)e^{-i\tilde{\omega}_3 t}$, with

$$\begin{aligned}\rho^{(2)}_{nm}(\mathbf{k};\omega_3) &= \frac{e}{i\hbar}\frac{1}{\omega_{nm\mathbf{k}}-\tilde{\omega}_3}\Big[-(B^c_{nm}(\mathbf{k},\omega_\beta)_{;k^b} \\ &\quad + i\sum_{\ell}\Big(r^b_{n\ell}B^c_{\ell m}(\mathbf{k},\omega_\beta) - B^c_{n\ell}(\mathbf{k},\omega_\beta)r^b_{\ell m}\Big)\Big]E^b_{\omega_\alpha}E^c_{\omega_\beta}\end{aligned} \tag{54}$$

where $\tilde{\omega}_3 = \tilde{\omega}_\alpha + \tilde{\omega}_\beta$ and $\mathbf{E}_{\omega_i}$ is the amplitude of the perturbing field with $\omega_i$ for $i = \alpha, \beta$. We use Eq. 54 in section 5.

## 4. Layered Current Density

In this section, we derive the expressions for the macroscopic current density of a given layer in the unit cell of the system. The approach we use to study the surface of a semi-infinite semiconductor crystal is as follows. Instead of using a semi-infinite system, we replace it by a slab (see Fig. 1). The slab consists of two surfaces, say the front and the back surface, and in between these two surfaces the bulk of the system. In general the surface of a crystal reconstructs as the atoms move to find equilibrium positions. This is due to the fact that the otherwise balanced forces are disrupted when the surface atoms do not find any more their bulk partner atoms, since these, by definition, are absent above (below) the front (back) surface of the slab. Therefore, to take the reconstruction into account, by surface we really mean the true surface that consists of the very first relaxed layer of atoms, and some of the sub-true-surface relaxed atomic layers. Since the front and the back surfaces of the slab are usually identical, the total slab is centrosymmetric. This fact (see Sec. 4), will imply $\chi^{slab}_{abc} = 0$, and thus we must device a way in which this artifact of a centrosymmetric

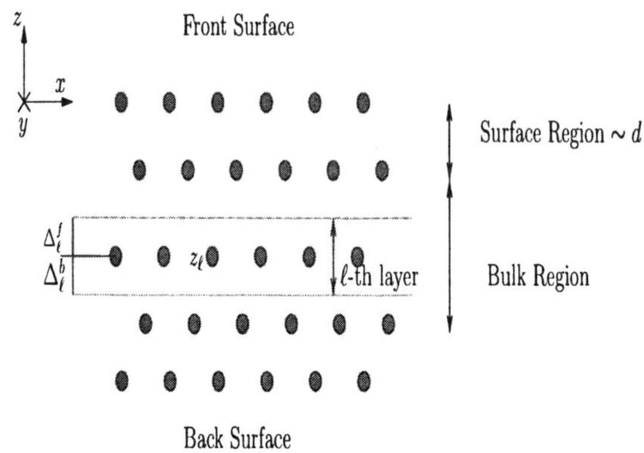


Figure 1. We show a sketch of the slab, where the small circles represent the atoms. See the text for the details.

slab is bypassed in order to have a finite $\chi^s_{abc}$ representative of the surface. Even if the front and back surfaces of the slab are different, thus breaking the centrosymmetry and therefore giving an overall $\chi^{slab}_{abc} \neq 0$, we need a procedure to extract the front surface $\chi^f_{abc}$ and the back surface $\chi^b_{abc}$ from the slab non-linear susceptibility $\chi^{slab}_{abc}$.

A convenient way to accomplish the separation of the SH signal of either surface is to introduce the so called "cut function", $S(z)$, which is usually taken to be unity over one half of the slab, and zero over the other half. In this case, $S(z)$ will give the contribution of the side of the slab for which $S(z) = 1$. However, we can generalize this simple choice for $S(z)$, by a top-hat cut function $S_\ell(z)$, that selects a given layer,

$$S_\ell(z) = \Theta(z - z_\ell + \Delta^b_\ell)\Theta(z_\ell - z + \Delta^f_\ell), \tag{55}$$

where $\Theta$ is the Heaviside function. Here, $\Delta^{f/b}_\ell$ is the distance that the $\ell$-th layer extends towards the front ($f$) or back ($b$) from its $z_\ell$ position. Thus $\Delta^f_\ell + \Delta^b_\ell$ is the thickness of layer $\ell$ (see Fig. 1).

Now, we show how this "cut function" $S_\ell(z)$ is introduced in the calculation of $\chi_{ijl}$. The microscopic current density is given by

$$\mathbf{j}(\mathbf{r}, t) = eTr(\hat{\mathbf{j}}(\mathbf{r})\hat{\rho}(t)), \tag{56}$$

where the operator for the electron's current is

$$\hat{\mathbf{j}}(\mathbf{r}) = \frac{1}{2}\left(\hat{\mathbf{v}}|\mathbf{r}\rangle\langle\mathbf{r}| + |\mathbf{r}\rangle\langle\mathbf{r}|\hat{\mathbf{v}}\right), \tag{57}$$

where $\hat{\mathbf{v}}$ is the electron's velocity operator to be dealt with below, and $Tr$ denotes the trace. We define $\hat{\mu} \equiv |\mathbf{r}\rangle\langle\mathbf{r}|$ and use the cyclic invariance of the trace to write

$$
\begin{aligned}
Tr(\hat{\mathbf{j}}(\mathbf{r})\hat{\rho}(t) &= Tr(\hat{\rho}(t)\hat{\mathbf{j}}(\mathbf{r})) = \frac{1}{2}\left(Tr(\hat{\rho}\hat{\mathbf{v}}\hat{\mu}) + Tr(\hat{\rho}\hat{\mu}\hat{\mathbf{v}})\right) \\
&= \frac{1}{2}\sum_{n\mathbf{k}}\left(\langle n\mathbf{k}|\hat{\rho}\hat{\mathbf{v}}\hat{\mu}|n\mathbf{k}\rangle + \langle n\mathbf{k}|\hat{\rho}\hat{\mu}\hat{\mathbf{v}}|n\mathbf{k}\rangle\right) \\
&= \frac{1}{2}\sum_{nm\mathbf{k}}\langle n\mathbf{k}|\hat{\rho}|m\mathbf{k}\rangle\left(\langle m\mathbf{k}|\hat{\mathbf{v}}|\mathbf{r}\rangle\langle \mathbf{r}|n\mathbf{k}\rangle + \langle m\mathbf{k}|\mathbf{r}\rangle\langle \mathbf{r}|\hat{\mathbf{v}}|n\mathbf{k}\rangle\right) \\
\mathbf{j}(\mathbf{r},t) &= e\sum_{nm\mathbf{k}}\rho_{nm}(\mathbf{k};t)\mathbf{j}_{mn}(\mathbf{k};\mathbf{r}),
\end{aligned}
\tag{58}
$$

where

$$
\mathbf{j}_{mn}(\mathbf{k};\mathbf{r}) = \frac{1}{2}\left(\langle m\mathbf{k}|\hat{\mathbf{v}}|\mathbf{r}\rangle\langle \mathbf{r}|n\mathbf{k}\rangle + \langle m\mathbf{k}|\mathbf{r}\rangle\langle \mathbf{r}|\hat{\mathbf{v}}|n\mathbf{k}\rangle\right), \tag{59}
$$

are the matrix elements of the microscopic current operator, and we have used the fact that the matrix elements between states $|n\mathbf{k}\rangle$ are diagonal in $\mathbf{k}$, i.e. proportional to $\delta(\mathbf{k}-\mathbf{k}')$.

Integrating the microscopic current $\mathbf{j}(\mathbf{r},t)$ over the entire slab gives the total macroscopic current density, however, if we want the contribution from only one region of the unit cell towards the total current, we can integrate $\mathbf{j}(\mathbf{r},t)$ over the desired region. The contribution to the current density from the $\ell$-th layer of the slab is given by

$$
\frac{1}{\Omega}\int d^3r\, S_\ell(z)\,\mathbf{j}(\mathbf{r},t) \equiv \mathbf{J}^{(\ell)}(t), \tag{60}
$$

where $\mathbf{J}^{(\ell)}(t)$ is the microscopic current in the $\ell$-th layer. Therefore we define

$$
\boldsymbol{\mathcal{V}}^{(\ell)}_{mn}(\mathbf{k}) \equiv \frac{1}{\Omega}\int d^3r\, S_\ell(z)\,\mathbf{j}_{mn}(\mathbf{k};\mathbf{r}), \tag{61}
$$

to write

$$
J^{(N,\ell)}_a(t) = e\sum_{mn\mathbf{k}}\mathcal{V}^{a(\ell)}_{mn}(\mathbf{k})\rho^{(N)}_{nm}(\mathbf{k};t), \tag{62}
$$

as the induced macroscopic current, to order $N$-th in the external perturbation, of the $\ell$-th layer. The matrix elements of the density operator for $N=1,2$ are given by Eqs. 50 and 54, respectively. Also, the roman superindices $a,b,c$ denote Cartesian components.

We proceed to give an explicit expression for $\mathcal{V}^{a(\ell)}_{mn}(\mathbf{k})$, for which we should work with the velocity operator, that is given by

$$
\begin{aligned}
i\hbar\hat{\mathbf{v}} &= [\hat{\mathbf{r}}, \hat{H}_0] \\
&= [\hat{\mathbf{r}}, \frac{\hat{\mathbf{p}}^2}{2m} + \hat{V}(\mathbf{r}) + \hat{v}(\mathbf{r},\hat{\mathbf{p}})] \approx [\hat{\mathbf{r}}, \frac{\hat{\mathbf{p}}^2}{2m}] = i\hbar\frac{\hat{\mathbf{p}}}{m},
\end{aligned}
\tag{63}
$$

where the possible contribution of the non-local pseudopotential $\hat{v}(\mathbf{r},\hat{\mathbf{p}})$ is neglected. Now, from above equation,

$$m\hat{\mathbf{v}} \approx \hat{\mathbf{p}} = -i\hbar\boldsymbol{\nabla}, \tag{64}$$

is the explicit functional form of the velocity or momentum operator. From Eq. 59, we need

$$\langle \mathbf{r}|\hat{\mathbf{v}}|n\mathbf{k}\rangle = \int d^3r' \langle \mathbf{r}|\hat{\mathbf{v}}|\mathbf{r}'\rangle\langle \mathbf{r}'|n\mathbf{k}\rangle \approx \frac{1}{m}\hat{\mathbf{p}}\psi_{n\mathbf{k}}(\mathbf{r}), \tag{65}$$

where we used

$$\langle \mathbf{r}|\hat{v^x}|\mathbf{r}'\rangle \approx \frac{1}{m}\langle \mathbf{r}|\hat{p^x}|\mathbf{r}'\rangle = \delta(y-y')\delta(z-z')\left(-i\hbar\frac{\partial}{\partial x}\delta(x-x')\right), \tag{66}$$

with similar results for the $y$ and $z$ Cartesian directions. Now, from Eqs. 61 and 59 we obtain

$$\boldsymbol{\mathcal{V}}^{(\ell)}_{mn}(\mathbf{k}) = \frac{1}{2}\int \mathrm{d}^3r\, S_\ell(z)\Big[\langle m\mathbf{k}|\mathbf{v}|\mathbf{r}\rangle\langle \mathbf{r}|n\mathbf{k}\rangle + \langle m\mathbf{k}|\mathbf{r}\rangle\langle \mathbf{r}|\mathbf{v}|n\mathbf{k}\rangle\Big], \tag{67}$$

and using Eq. 65, we can write, for any function $S(z)$ used to identify the response from a region of the slab, that

$$\begin{aligned}
\boldsymbol{\mathcal{V}}_{mn}(\mathbf{k}) &\approx \frac{1}{2m}\int d^3r S(z)\Big[\psi_{n\mathbf{k}}(\mathbf{r})\hat{\mathbf{p}}^*\psi^*_{m\mathbf{k}}(\mathbf{r}) + \psi^*_{m\mathbf{k}}(\mathbf{r})\hat{\mathbf{p}}\psi_{n\mathbf{k}}(\mathbf{r})\Big], && (68)\\
&= \frac{1}{m}\int d^3r\psi^*_{m\mathbf{k}}(\mathbf{r})\left[\frac{S(z)\mathbf{p}+\mathbf{p}S(z)}{2}\right]\psi_{n\mathbf{k}}(\mathbf{r}), && (69)\\
&= \frac{1}{m}\int d^3r\psi^*_{m\mathbf{k}}(\mathbf{r})\hat{\boldsymbol{\mathcal{P}}}\psi_{n\mathbf{k}}(\mathbf{r}) \equiv \frac{1}{m}\boldsymbol{\mathcal{P}}_{mn}(\mathbf{k}). && (70)
\end{aligned}$$

Here an integration by parts is performed on the first term of the right hand side of Eq. 68; since the $e^{-i\mathbf{k}\cdot\mathbf{r}}\psi_{n\mathbf{k}}(\mathbf{r})$ are periodic over the unit cell, the surface term vanishes. From Eqs. 68 we see that the replacement

$$\hat{\mathbf{p}} \to \hat{\boldsymbol{\mathcal{P}}} = \left[\frac{S(z)\hat{\mathbf{p}}+\hat{\mathbf{p}}S(z)}{2}\right], \tag{71}$$

is what it takes to change the momentum operator of the electron, $\hat{\mathbf{p}}$, to the new momentum operator $\hat{\boldsymbol{\mathcal{P}}}$ that implicitly takes into account the contribution of the region of the slab given by $S(z)$. Note that $\hat{\boldsymbol{\mathcal{P}}}$ is properly symmetrized.

Finally, the Fourier component of macroscopic current of Eq. 62 is given by

$$J^{(N,\ell)}_a(\omega_3) = \frac{e}{m}\sum_{mn\mathbf{k}} \mathcal{P}^{a(\ell)}_{mn}(\mathbf{k})\rho^{(N)}_{nm}(\mathbf{k};\omega_3), \tag{72}$$

where the non-local contribution of $H_0$ is neglected, and from Eq. 69

$$\mathcal{P}^{a(\ell)}_{mn} = \int d^3r\psi^*_{m\mathbf{k}}(\mathbf{r})\left[\frac{S_\ell(z)p^a + p^a S_\ell(z)}{2}\right]\psi_{n\mathbf{k}}(\mathbf{r}). \tag{73}$$

Actually, to limit the response to one surface, the Eq. 71 was proposed in Ref. [15], and latter used in Refs. [16] and [17] in the context of SHG. Then, the layer-by-layer analysis of Refs. [18] and [19] actually used Eq. 55 thus limiting the current response to a particular layer of the slab, and used it to obtain the anisotropic linear optical response of semiconductor surfaces. However, the first formal derivation of this scheme is presented in Ref. [20] for the linear optical response, and here for the non-linear optical response of semiconductors.

## 5. Non-linear Susceptibility

In this section we obtain the expressions for the non-linear surface susceptibility tensor to second order in the perturbing fields. We start with the non-linear polarization $\mathbf{P}$ written as

$$\begin{aligned} P_a(\omega_3) &= \chi_{abc}(-\omega_3;\omega_1,\omega_2)E_b(\omega_1)E_c(\omega_2) \\ &\quad+\chi_{abcl}(-\omega_3;\omega_1,\omega_2)E_b(\omega_1)\nabla_c E_l(\omega_2)+\cdots, \end{aligned} \tag{74}$$

where $\chi_{abc}$ and $\chi_{abcl}$, correspond to the dipolar and quadrupolar susceptibilities, respectively, and the sum continues with higher multipolar terms. If we consider a semi-infinite system with a centrosymmetric bulk, above equation splits, due to symmetry considerations alone, into two contributions, one from the surface of the system and the other from the bulk of the system. Indeed, let's take

$$P_a(\mathbf{r}) = \chi_{abc}E_b(\mathbf{r})E_c(\mathbf{r}) + \chi_{abcl}E_b(\mathbf{r})\frac{\partial}{\partial \mathbf{r}_c}E_l(\mathbf{r}) + \cdots, \tag{75}$$

as the polarization with respect to the original coordinate system, and

$$\begin{aligned} P_a(-\mathbf{r}) &= \chi_{abc}E_b(-\mathbf{r})E_c(-\mathbf{r}) \\ &\quad+\chi_{abcl}E_b(-\mathbf{r})\frac{\partial}{\partial(-\mathbf{r}_c)}E_l(-\mathbf{r})+\cdots, \end{aligned} \tag{76}$$

as the polarization in the coordinate system where inversion is taken, i.e. $\mathbf{r}\to-\mathbf{r}$. Note that we have kept the same susceptibility tensors, since as the system is centrosymmetric, they must be invariant under $\mathbf{r}\to-\mathbf{r}$. Recalling that $\mathbf{P}(\mathbf{r})$ and $\mathbf{E}(\mathbf{r})$, are polar vectors,[21] we have that Eq. 76 reduces to

$$-P_a(\mathbf{r}) = \chi_{abc}(-E_b(\mathbf{r}))(-E_c(\mathbf{r})) - \chi_{abcl}(-E_b(\mathbf{r}))(-\frac{\partial}{\partial \mathbf{r}_c})(-E_l(\mathbf{r})) + \cdots,$$

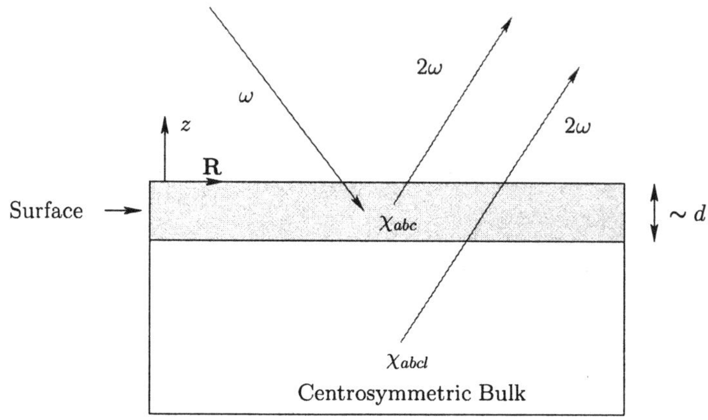


Figure 2. (color online) We show a sketch of the semi-infinite system with a centrosymmetric bulk. The surface region is of width $\sim d$. The incoming photon of frequency $\omega$ is represented by a downward red arrow, whereas both the surface and bulk created second harmonic photons of frequency $2\omega$ are represented by an upward green arrow. The red color suggests an infrared incoming photon whose second harmonic generated photon is in the green. The dipolar, $\chi_{abc}$, and quadrupolar, $\chi_{abcl}$, susceptibility tensors are shown in the regions where they are different from zero. The axis are also shown, with $z$ perpendicular to the surface and $\mathbf{R}$ parallel to it.

$$P_a(\mathbf{r}) = -\chi_{abc}E_b(\mathbf{r})E_c(\mathbf{r}) + \chi_{abcl}E_b(\mathbf{r})\frac{\partial}{\partial \mathbf{r}_c}E_l(\mathbf{r}) + \cdots, \tag{77}$$

that when compared with Eq. 75 leads to the conclusion that

$$\chi_{abc} = 0 \qquad \text{for a centrosymmetric bulk.} \tag{78}$$

Therefore, if we move to the surface of the semi-infinite system, the assumption of centrosymmetry necessarily breaks down, and there is no restriction in $\chi_{abc}$. Thus, we conclude that the leading term of the polarization in a surface region is given by

$$\begin{aligned}\int d\mathbf{R}\int dz P_a(\mathbf{R},z) &\approx \mathcal{S}dP_a \\ &= \mathcal{S}P_a^s \\ &= \chi_{abc}E_bE_c,\end{aligned} \tag{79}$$

where $\mathbf{R}$ is a vector parallel to the surface which is perpendicular to $z$, $\mathcal{S}$ is the surface area of the unit cell that characterizes the surface of the system, and $d$ is the surface region from which the dipolar signal of $\mathbf{P}$ is different from zero (see Fig. 2). Also, $d\mathbf{P} \equiv \mathbf{P}^{(s)}$ is the surface SH polarization, given by

$$P_a^s = \frac{1}{\mathcal{S}}\chi_{abc}E_bE_c = \chi^s_{abc}E_bE_c, \tag{80}$$

with $\chi^s_{abc} = \chi_{abc}/\mathcal{S}$ the surface non-linear susceptibility. On the other hand,

$$P^b_a(\mathbf{r}) = \chi_{abcl} E_b(\mathbf{r}) \nabla_c E_l(\mathbf{r}), \tag{81}$$

gives the bulk polarization. Immediately we see that the surface polarization is of dipolar order, whereas the bulk polarization is of quadrupolar order, and that the rank of the susceptibility tensors is 3 for the surface, i.e. $\chi_{abc}$, and 4 for the bulk, i.e. $\chi_{abcl}$. Although the bulk generated SH is in itself a very important optical phenomena, in here we concentrate only in the surface generated SH. Indeed, in centrosymmetric systems for which the quadrupolar bulk response is much smaller that the dipolar surface response, SH is readily used as a very useful and powerful optical surface probe.[1]

To calculate $\chi^s_{abc}$, we start from the basic relation, $\mathbf{J} = d\mathbf{P}/dt$ with $\mathbf{J}$ the current calculated in Sec. 4, and from Eq. 72 we obtain

$$J^{(2,\ell)}_a(\omega_3) = -i\omega_3 P_a(\omega_3) = \frac{e}{m} \sum_{mn\mathbf{k}} \mathcal{P}^{a(\ell)}_{mn}(\mathbf{k}) \rho^{(2)}_{nm}(\mathbf{k};\omega_3), \tag{82}$$

which upon using Eqs. 54 and 80 leads to

$$\begin{aligned}
\chi^{s(\ell)}_{abc}(-\omega_3;\omega_1,\omega_2) &= \frac{ie}{mE^b_1 E^c_2 \mathcal{S}\omega_3} \sum_{mn\mathbf{k}} \mathcal{P}^{a(\ell)}_{mn}(\mathbf{k}) \rho^{(2)}_{nm}(\mathbf{k};\omega_3) \\
&= \frac{e^2}{\mathcal{S}m\hbar\omega_3} \sum_{mn\mathbf{k}} \frac{\mathcal{P}^{a(\ell)}_{mn}(\mathbf{k})}{\omega_{nm\mathbf{k}} - \tilde{\omega}_3} \Bigg[ - (B^c_{nm}(\mathbf{k},\omega_\beta))_{;k^b} \\
&\quad + i \sum_\ell \left( r^b_{n\ell} B^c_{\ell m}(\mathbf{k},\omega_\beta) - B^c_{n\ell}(\mathbf{k},\omega_\beta) r^b_{\ell m} \right) \Bigg],
\end{aligned} \tag{83}$$

which gives the surface susceptibility of layer $\ell$-th. As can be seen from Eq. (54), $\chi^{s(\ell)}_{abc}$ can be split into two terms, one coming from the first term of Eq. (54) and the other from the second term of the same equation. Then we have, after substituting Eq. 50, that

$$\chi^{s(\ell)}_{i,abc} = -\frac{e^3}{m\hbar^2\omega_3} \sum_{mn\mathbf{k}} \frac{\mathcal{P}^{a(\ell)}_{mn}}{\omega_{nm} - \omega_3} \left( \frac{f_{mn} r^b_{nm}}{\omega_{nm} - \omega_\beta} \right)_{;k^c}, \tag{84}$$

and

$$\chi^{s(\ell)}_{e,abc} = \frac{ie^3}{m\hbar^2\omega_3} \sum_{\ell mn\mathbf{k}} \frac{\mathcal{P}^{a(\ell)}_{mn}}{\omega_{nm} - \omega_3} \left( \frac{r^c_{n\ell} r^b_{\ell m} f_{m\ell}}{\omega_{\ell m} - \omega_\beta} - \frac{r^b_{n\ell} r^c_{\ell m} f_{\ell n}}{\omega_{n\ell} - \omega_\beta} \right), \tag{85}$$

where $\chi^{s(\ell)}_i$ is related to intraband transitions and $\chi^{s(\ell)}_e$ to interband transitions. We mention that Eq. (84) and Eq. (85) need to be symmetrized for intrinsic permutation symmetry, i.e. $\chi^{abc}(-\omega_3;\omega_1,\omega_2) = \chi^{acb}(-\omega_3;\omega_2,\omega_1)$,[22] and that for SHG $\omega_1 = \omega_2 = \omega$ and $\omega_3 = 2\omega$.

The generalized derivative in Eq. (84) is obtained from the chain rule as

$$\left(\frac{f_{mn}r^b_{nm}}{\omega_{nm}-\omega_2}\right)_{;k^c} = \frac{f_{mn}}{\omega_{nm}-\omega}\left(r^b_{nm}\right)_{;k^c} - \frac{f_{mn}r^b_{nm}}{(\omega_{nm}-\omega)^2}\left(\omega_{nm}\right)_{;k^c}, \tag{86}$$

here $(\omega_{nm})_{;k^a} = (\omega_n)_{;k^a} - (\omega_m)_{;k^a}$. In the appendices we show that

$$(\omega_{nm})_{;k^c} = \frac{p^c_{nn} - p^c_{mm}}{m} \equiv \Delta^c_{nm}, \tag{87}$$

and that

$$(r^b_{nm})_{;k^c} = \frac{r^c_{nm}\Delta^b_{mn} + r^b_{nm}\Delta^c_{mn}}{\omega_{nm}} + \frac{i}{\omega_{nm}}\sum_{\ell}\left(\omega_{\ell m}r^c_{n\ell}r^b_{\ell m} - \omega_{n\ell}r^b_{n\ell}r^c_{\ell m}\right). \tag{88}$$

Above formulas give a complete set of relationships in order to calculate the nonlinear susceptibility of any given layer $\ell$ as $\boldsymbol{\chi}^{s(\ell)} = \boldsymbol{\chi}^{s(\ell)}_e + \boldsymbol{\chi}^{s(\ell)}_i$. Then, we can calculate the surface susceptibility as

$$\chi^s_{abc}(2\omega) \equiv \sum_{\ell_0}^{\ell_d} \chi^{(\ell)}_{abc}(2\omega), \tag{89}$$

where $\ell_0$ represents the first layer right at the surface, and $\ell_d$ the layer at a distance $\sim d$ from the surface (see Fig. 2). Of course we can use Eq. 89 for either the front or the back surface. Likewise

$$\chi^{(\ell_f)}_{abc}(2\omega) \equiv \sum_{\ell_d}^{\ell_f} \chi^{(\ell)}_{abc}(2\omega), \tag{90}$$

is a dipolar bulk susceptibility, with the property that,

$$\chi^{(\ell_f)}_{abc}(2\omega) \overset{\ell_f\to\ell_b}{=} 0, \tag{91}$$

where $\ell_b$ is a bulk layer such that the bulk centrosymmetry is fully stablished and the dipolar non-linear susceptibility is identically zero, in accordance with Eq. 78. We remark that $\ell_d$ is not universal, and $\ell_b$ should be found according to Eq. 91.

In the next section we show that the longitudinal and transverse gauge formally give the same result for any order $N$, and present a simple relationship that can be checked numerically.

## 6. Gauge Invariance

We present a general procedure to stablish the equivalence between the longitudinal and the transverse gauge. In the transverse gauge the interaction hamiltonian is given by $\hat{H}' = (-e/mc)\hat{\mathbf{p}}\cdot\mathbf{A}$. Within the long wavelength approximation, **A** is constant through out space, and since we are working in the semiclassical approximation, **A** is not an operator. From Eq. 28 we can get

$$i\hbar(\hat{\rho}_{I,T}(t) - \rho_0) = \frac{-e}{mc}\int_{-\infty}^{t} dt'[\hat{\mathbf{p}}_I(t')\cdot\mathbf{A}(t'), \hat{\rho}_{I,T}(t')], \tag{92}$$

where $T$ denotes the transverse gauge. An integration by parts gives

$$\begin{aligned} i\hbar(\hat{\rho}_{I,T}(t) - \rho_0) = \frac{-e}{c}[\hat{\mathbf{r}}_I(t)\cdot\mathbf{A}(t), \hat{\rho}_{I,T}(t)] - e\int_{-\infty}^{t} dt'[\hat{\mathbf{r}}_I\cdot\mathbf{E}, \hat{\rho}_{I,T}(t')] \\ -\frac{e}{c}\int_{-\infty}^{t} dt'[\hat{\mathbf{r}}_I\cdot\mathbf{A}, \dot{\hat{\rho}}_{I,T}(t')]. \end{aligned} \tag{93}$$

To first order we get

$$\hat{\rho}^{(1)}_{I,T}(t) = \hat{\rho}^{(1)}_{I,L}(t) - \frac{e}{i\hbar c}[\hat{\mathbf{r}}_I(t)\cdot\mathbf{A}(t), \rho_0], \tag{94}$$

where $L$ denotes the longitudinal gauge. With the velocity operator (T-gauge) $\mathbf{v} = \mathbf{p}/m - e\mathbf{A}/mc$, Eq. (72) gives $\chi^{(1)}_T = \chi^{(1)}_L + S^{(1)}$, where for a harmonic perturbation,

$$S^{(1)} = \frac{-ic^2}{\omega^2 m\hbar}\sum_{mn}(C_1)_{mn}(\rho_0)_{nm}, \tag{95}$$

with $C_1 = [r^a, p^b] - i\hbar\delta^{ab}$. Therefore, since formally $[r^a, p^b] = i\hbar\delta^{ab}$, $S^{(1)} = 0$, and $\chi^{(1)}_T = \chi^{(1)}_L$, which shows gauge invariance. From Eq. (94) one can show to any order that $\chi^{(N)}_T = \chi^{(N)}_L + S^{(N)}$, where again $S^{(N)} = 0$, due to commutator identities. Therefore, the gauge invariance relays on the fulfillment of the commutator relationships, and thus one has to check this for the particular system in question. As an example, we can take matrix elements of the simple commutator

$$([r^a, p^b])_{nn} = \sum_m \left(r^a_{nm}p^b_{mn} - p^b_{nm}r^a_{mn}\right) = \sum_m \frac{p^a_{nm}p^b_{mn} - p^b_{nm}p^a_{mn}}{im\omega_{nm}}, \tag{96}$$

to obtain

$$\sum_m \frac{p^a_{nm}p^b_{mn} - p^b_{nm}p^a_{mn}}{\omega_{nm}} = m\hbar\delta^{ab}, \qquad \forall\, n \neq m \tag{97}$$

which is a relationship that can be numerically verified.

## 7. SHG Radiation

In this section we derive the formulas required for the calculation of the SHG yield, defined by

$$R(\omega) = I(2\omega)/I^2(\omega), \tag{98}$$

with the intensity

$$I(\omega) = c/8\pi|E(\omega)|^2. \tag{99}$$

There are several ways to calculate $R$, one of which is the procedure followed by Cini.[23] This approach calculates the non-linear susceptibility and at the same time the radiated fields. However, we present an alternative derivation based in the work of Mizrahi and Sipe,[24] since the derivation of the so called three-layer-model is straightforward. Within our level of approximation this is the best model that we can use. In this scheme, we assume that the SH conversion takes place in a thin layer, just below the surface, that is characterized by a surface dielectric function $\epsilon_\ell(\omega)$. This layer is below vacuum and sits on top of the bulk characterized by $\epsilon_b(\omega)$ (see Fig. 3). The non-linear polarization immersed in the thin layer, will radiate an electric field directly into vacuum and also into the bulk. This bulk directed field, will be reflected back into vacuum. Thus, the total field radiated into vacuum will be the sum of these two contributions (see Fig. 3). We decompose the field into $s$ and $p$ polarizations, then the electric field radiated by a polarization sheet of the form given by Eq. 80, $\mathcal{P}_i = \chi_{ijk} E_j(\omega) E_k(\omega)$,[b] is given by,[24]

$$(E_{p\pm}, E_s) = (\frac{2\pi i\tilde{\omega}^2}{w}\,\hat{\mathbf{p}}_\pm \cdot \boldsymbol{\mathcal{P}}, \frac{2\pi i\tilde{\omega}^2}{w}\,\hat{\mathbf{s}} \cdot \boldsymbol{\mathcal{P}}), \tag{100}$$

where $\hat{\mathbf{s}}$ and $\hat{\mathbf{p}}_\pm$ are the unitary vectors for $s$ and $p$ polarization, respectively, and the $\pm$ refers to upward $(+)$ or downward $(-)$ direction of propagation. Also, $\tilde{\omega} = \omega/c$ and $w = \tilde{\omega} k_z$, with

$$k_z(\omega) = \sqrt{\epsilon(\omega) - \sin^2\theta}, \tag{101}$$

and $\hat{\mathbf{p}}_\pm = \mathbf{p}_\pm/\sqrt{\epsilon}$, with

$$\mathbf{p}_\pm = \mp k_z \hat{\mathbf{x}} - \sin\theta\hat{\mathbf{z}}. \tag{102}$$

In the above equations $z$ is the direction perpendicular to the surface that points towards the bulk, $x$ is parallel to the surface, and $\theta$ is the angle of incidence, where the plane of incidence is chosen as the $xz$ plane (see

---

[b] For convenience of notation we now use subscripts $i, j, k$ as Cartesian indices.

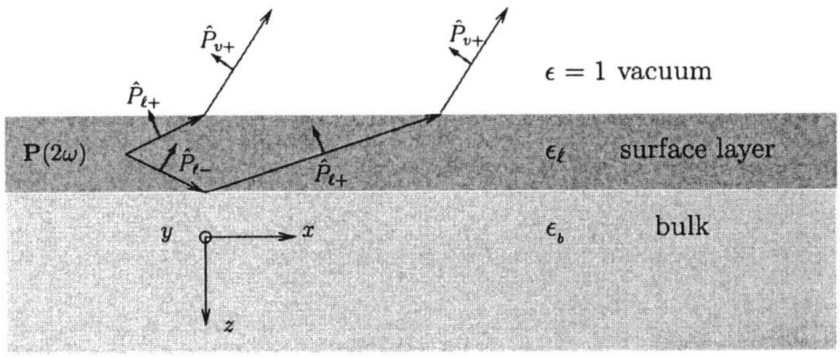


Figure 3. Sketch of the three layer model for SHG. Vacuum is on top with $\epsilon = 1$, the layer with non-linear polarization $\mathbf{P}$ is characterized with $\epsilon_\ell(\omega)$ and the bulk with $\epsilon_b(\omega)$. In the dipolar approximation the bulk does not radiate SHG. The thin arrows are along the direction of propagation, and the unit vectors for $p$-polarization are denoted with thick arrows (capital letters denote SH components). The unit vector for $s$-polarization points along $y$ (out of the page).

Fig. 3), thus $\hat{\mathbf{s}} = \hat{\mathbf{y}}$. The function $k_z(\omega)$ is the projection of the wave vector perpendicular to the surface. As we see from Fig. 3, the SH field is refracted at the layer-vacuum interface ($\ell v$), and reflected from the layer-bulk ($\ell b$) interface, thus we can define the transmission, $\mathbf{T}$, and reflection, $\mathbf{R}$, tensors as,

$$\mathbf{T}_{\ell v} = \hat{\mathbf{s}} T_s^{\ell v} \hat{\mathbf{s}} + \hat{\mathbf{P}}_{v+} \tilde{T}_p^{\ell v} \hat{\mathbf{P}}_{\ell +}, \tag{103}$$

and

$$\mathbf{R}_{\ell b} = \hat{\mathbf{s}} R_s^{\ell b} \hat{\mathbf{s}} + \hat{\mathbf{P}}_{\ell +} R_p^{\ell b} \hat{\mathbf{P}}_{\ell -}, \tag{104}$$

where variables in capital letters are evaluated at the harmonic frequency $2\omega$. Notice that since $\hat{\mathbf{s}}$ is independent of $\omega$, then $\hat{\mathbf{s}} = \hat{\mathbf{s}}$. The Fresnel factors, $T_i$, $R_i$, and $\tilde{T}_p$, for $i = s, p$ polarization, are evaluated at the appropriate interface $\ell v$ or $\ell b$, and will be given below. The extra subscript in $\hat{\mathbf{P}}$ denotes the corresponding dielectric function to be used in its evaluation, i.e. $\epsilon_v = 1$ for vacuum ($v$), $\epsilon_\ell$ for the layer ($\ell$), and $\epsilon_b$ for the bulk ($b$). Therefore, the total radiated field at $2\omega$ is

$$\begin{aligned}\mathbf{E}(2\omega) &= E_s(2\omega)\left(\mathbf{T}_{\ell v} + \mathbf{T}_{\ell v} \cdot \mathbf{R}_{\ell b}\right) \cdot \hat{\mathbf{s}} \\ &+ E_{p+}(2\omega)\mathbf{T}_{\ell v} \cdot \hat{\mathbf{P}}_{\ell +} + E_{p-}(2\omega)\mathbf{T}_{\ell v} \cdot \mathbf{R}_{\ell b} \cdot \hat{\mathbf{P}}_{\ell -}.\end{aligned} \tag{105}$$

The first term is the transmitted $s$-polarized field, the second one is the reflected and then transmitted $s$-polarized field and the third and fourth terms are the equivalent fields for $p$-polarization. The transmission is from the layer into vacuum, and the reflection between the layer and the bulk. After some simple algebra, we obtain

$$\mathbf{E}(2\omega) = \frac{4\pi i\tilde{\omega}}{K_{z\ell}} \mathbf{H} \cdot \boldsymbol{\mathcal{P}}, \tag{106}$$

where,

$$\mathbf{H} = \hat{\mathbf{s}}\, T_s^{\ell v} \left(1 + R_s^{\ell b}\right) \hat{\mathbf{s}} + \hat{\mathbf{P}}_{v+} \tilde{T}_p^{\ell v} \left(\hat{\mathbf{P}}_{\ell+} + R_p^{\ell b} \hat{\mathbf{P}}_{\ell-}\right). \tag{107}$$

The magnitude of the radiated field is given by $E(2\omega) = \hat{\mathbf{e}}^{out} \cdot \mathbf{E}(2\omega)$, where $\hat{\mathbf{e}}^{out}$ is the polarization vector of the radiated field, for instance $\hat{\mathbf{s}}$ or $\hat{\mathbf{P}}_{v+}$. Then we write

$$E(2\omega) = \frac{4\pi i\omega}{c} \mathbf{e}^{2\omega} \cdot \boldsymbol{\mathcal{P}}. \tag{108}$$

Using the above equations and the following simple relationships between $T$ and $R$,

$$T_s^{\ell v} = \frac{K_{z\ell}}{\cos\theta} T_s^{v\ell}, \qquad \tilde{T}_p^{\ell v} = \frac{\sqrt{\epsilon_\ell(2\omega)} K_{z\ell}}{\cos\theta} T_p^{v\ell}, \tag{109}$$

$$1 - R_p^{\ell b} = \frac{\epsilon_\ell(2\omega) K_{zb}}{K_{z\ell}} T_p^{\ell b}, \qquad 1 + R_p^{\ell b} = \epsilon_b(2\omega) T_p^{\ell b}, \tag{110}$$

we obtain

$$\mathbf{e}^{2\omega} = \frac{1}{\cos\theta} \hat{\mathbf{e}}^{out} \cdot \left[\hat{\mathbf{s}} T_s^{v\ell} T_s^{\ell b} \hat{\mathbf{s}} - \hat{\mathbf{P}}_{v+} T_p^{v\ell} T_p^{\ell b} \left(\epsilon_\ell(2\omega) K_{zb} \hat{\mathbf{x}} + \epsilon_b(2\omega) \sin\theta \hat{\mathbf{z}}\right)\right], \tag{111}$$

and then we write from Eq. (108)

$$E_s(2\omega) = \frac{4\pi i\omega}{c\cos\theta} T_s^{v\ell} T_s^{\ell b} \chi_{yij} E_i(\omega) E_j(\omega), \tag{112}$$

and

$$E_p(2\omega) = \frac{-4\pi i\omega}{c\cos\theta} T_p^{v\ell} T_p^{\ell b} \left[\epsilon_\ell(2\omega) K_{zb} \chi_{xij} + \epsilon_b(2\omega) \sin\theta \chi_{zij}\right] E_i(\omega) E_j(\omega). \tag{113}$$

As mentioned before $E_i(\omega)$ is the incident field given by the external field properly screened; then we have

$$\mathbf{E}_s(\omega) = E_o t_s^{v\ell} \left(1 + r_s^{\ell b}\right) \hat{\mathbf{y}}, \tag{114}$$

and

$$\mathbf{E}_p(\omega) = E_o \left[\tilde{t}_p^{v\ell} \left(1 - r_p^{\ell b}\right) \cos\theta_\ell \hat{\mathbf{x}} - \tilde{t}_p^{v\ell} \left(1 + r_p^{\ell b}\right) \sin\theta_\ell \hat{\mathbf{z}}\right], \tag{115}$$

where $E_o$ is the incoming amplitude and $\theta_\ell$ is the angle of refraction in the layer. Notice that the transmitted and reflected fields in the layer are taken into $\mathbf{E}_s$ and $\mathbf{E}_p$. From Eqs. (109-110) we get

$$\mathbf{E}_s(\omega) = E_o t_s^{v\ell} t_s^{\ell b} \hat{\mathbf{y}}, \tag{116}$$

and

$$\mathbf{E}_p(\omega) = E_o t_p^{v\ell} t_p^{\ell b} \left(\epsilon_\ell(\omega) k_{zb}\hat{\mathbf{x}} - \epsilon_b(\omega)\sin\theta\hat{\mathbf{z}}\right). \tag{117}$$

Using Eqs. (112), (113), (116), (117), into $R$, we finally write

$$R_{iF} = \frac{32\pi^3\omega^2}{(n_o e)^2 c^3 \cos^2\theta} \left|T_F^{v\ell} T_F^{\ell b} (t_i^{v\ell} t_i^{\ell b})^2 \, r_{iF}\right|^2, \tag{118}$$

where $i$ (lower case) stands for initial polarization and $F$ (upper case) stands for final polarization, with

$$r_{iP} = (K_{zb}\chi_{xjk} + \sin\theta\chi_{zjk})\, E_j^i E_k^i, \tag{119}$$

and

$$r_{iS} = \chi_{yjk} E_j^i E_k^i, \tag{120}$$

where from Eqs. (116-117), $\mathbf{E}^s = \hat{\mathbf{y}}$, and

$$\mathbf{E}^p = \epsilon_\ell(\omega) k_{zb}\hat{\mathbf{x}} - \epsilon_b(\omega)\sin\theta\hat{\mathbf{z}}. \tag{121}$$

The $n_o e$ factor in Eq. (118), with $n_o$ the electronic density, renders $\chi$ dimensionless. To complete the required formulas, we write down the Fresnel factors,

$$t_s^{v\ell} = \frac{2\cos\theta}{\cos\theta + k_{z\ell}}, \qquad t_p^{v\ell} = \frac{2\cos\theta}{\epsilon_\ell(\omega)\cos\theta + k_{z\ell}}, \tag{122}$$

$$t_s^{\ell b} = \frac{2k_{z\ell}}{k_{z\ell} + k_{zb}}, \qquad t_p^{\ell b} = \frac{2k_{z\ell}}{\epsilon_b(\omega) k_{z\ell} + \epsilon_s(\omega) k_{zb}}, \tag{123}$$

where the appropriate term $\sqrt{\epsilon(\omega)}$ from the usual definition of $t_p$,[21] has been taken out to give Eqs. (119) and (120). For a given surface symmetry and its corresponding non-zero tensor elements of $\chi_{ijk}$, Eq. (118) can be calculated explicitly through Eqs. (119) and (120) [6,25]. With the three-layer model we can get two opposite cases, one in which the SH conversion takes places in vacuum for which we simply put $\epsilon_\ell = 1$, and the other case where the layer is identical to the bulk, or $\epsilon_\ell = \epsilon_b$. The former case corresponds to no screening and the latter to the usual Fresnel screening.

## 8. Conclusions

We have presented a complete derivation of the required elements to calculate the surface SHG radiated from a semiconductor within the dipole approximation, and showed how to calculate the layer-by-layer contribution to the optical signal. We derived the nonlinear surface susceptibility tensor $\chi$ within the longitudinal gauge and thus we decomposed $\chi$ into intraband

and interband one-electron transitions. We showed that the longitudinal and transverse gauges give the same result, and a simple expression was presented in order to check the numerical accuracy of this equivalency. Also, we calculated the radiated efficiency $R$ within the three layer model. The combination of $\chi$ and $R$ allow us to study this fascinating surface optical phenomena.

**Acknowledgments**

We acknowledge the partial support from CONACyT-México under grant 48915-F.

## Appendix A.

We present some basic results needed in the derivation of the main results. The normalization of the states $\psi_{n\mathbf{q}}(\mathbf{r})$ are chosen such that

$$\psi_{m\mathbf{q}}(\mathbf{r}) = \left(\frac{\Omega}{8\pi^3}\right)^{\frac{1}{2}} u_{m\mathbf{q}}(\mathbf{r}) e^{i\mathbf{q}\cdot\mathbf{r}}, \tag{A.1}$$

and

$$\int_\Omega d^3r\, u^*_{n\mathbf{k}}(\mathbf{r}) u_{m\mathbf{q}}(\mathbf{r}) = \delta_{nm}\delta_{\mathbf{k},\mathbf{q}}, \tag{A.2}$$

where $\Omega$ is the volume is the unit cell and $\delta_{a,b}$ is the Kronecker delta that gives one if $a = b$ and zero otherwise. For box normalization, where we have $N$ unit cells in some volume $V = N\Omega$, this gives

$$\int_V d^3r\, \psi^*_{n\mathbf{k}}(\mathbf{r}) \psi_{m\mathbf{q}}(\mathbf{r}) = \frac{V}{8\pi^3}\delta_{nm}\delta_{\mathbf{k},\mathbf{q}}, \tag{A.3}$$

which lets us have in the limit of $N \to \infty$

$$\int d^3r\, \psi^*_{n\mathbf{k}}(\mathbf{r}) \psi_{m\mathbf{q}}(\mathbf{r}) = \delta_{nm}\delta(\mathbf{k} - \mathbf{q}), \tag{A.4}$$

for which the Kornecker-$\delta$ is replaced by

$$\delta_{\mathbf{k},\mathbf{q}} \to \frac{8\pi^3}{V}\delta(\mathbf{k} - \mathbf{q}), \tag{A.5}$$

and we recall that $\delta(x) = \delta(-x)$. Now, for any periodic function $f(\mathbf{r}) = f(\mathbf{r} + \mathbf{R})$ we have

$$\int d^3r\, e^{i(\mathbf{q}-\mathbf{k})\cdot\mathbf{r}} f(\mathbf{r}) = \sum_j^{\text{unit cells}} \int_\Omega d^3r\, e^{i(\mathbf{q}-\mathbf{k})\cdot(\mathbf{r}+\mathbf{R}_j)} f(\mathbf{r} + \mathbf{R}_j),$$

$$= \sum_{j}^{\text{unit cells}} \int_{\Omega} d^3r \, e^{i(\mathbf{q}-\mathbf{k})\cdot(\mathbf{r}+\mathbf{R}_j)} f(\mathbf{r}),$$

$$= \int_{\Omega} d^3r \, e^{i(\mathbf{q}-\mathbf{k})\cdot\mathbf{r}} f(\mathbf{r}) \sum_{j}^{\text{unit cells}} e^{i(\mathbf{q}-\mathbf{k})\cdot\mathbf{R}_j},$$

$$= \int_{\Omega} d^3r \, e^{i(\mathbf{q}-\mathbf{k})\cdot\mathbf{r}} f(\mathbf{r}) N \sum_{\mathbf{K}} \delta_{\mathbf{K},\mathbf{q}-\mathbf{k}},$$

$$= N \int_{\Omega} d^3r \, e^{i(\mathbf{q}-\mathbf{k})\cdot\mathbf{r}} f(\mathbf{r}) \delta_{\mathbf{0},\mathbf{q}-\mathbf{k}},$$

$$= N \delta_{\mathbf{q},\mathbf{k}} \int_{\Omega} d^3r \, f(\mathbf{r}),$$

$$= \frac{8\pi^3}{\Omega} \delta(\mathbf{q}-\mathbf{k}) \int_{\Omega} d^3r \, f(\mathbf{r}), \tag{A.6}$$

where we have assumed that $\mathbf{k}$ and $\mathbf{q}$ are restricted to the first Brillouin zone, and thus the reciprocal lattice vector $\mathbf{K} = 0$.

## Appendix B.

We obtain the generalized derivative $(\omega_n(\mathbf{k}))_{;\mathbf{k}}$. We start from

$$\langle n\mathbf{k}|\hat{H}_0|m\mathbf{k}'\rangle = \delta_{nm}\delta(\mathbf{k}-\mathbf{k}')\hbar\omega_m(\mathbf{k}), \tag{B.1}$$

then Eq. 26 gives

$$\begin{aligned}(H_{0,nm})_{;\mathbf{k}} &= \nabla_{\mathbf{k}} H_{0,nm}(\mathbf{k}) - iH_{0,nm}(\mathbf{k})\left(\boldsymbol{\xi}_{nn}(\mathbf{k}) - \boldsymbol{\xi}_{mm}(\mathbf{k})\right)\\ &= \delta_{nm}\hbar\nabla_{\mathbf{k}}\omega_m(\mathbf{k}),\end{aligned} \tag{B.2}$$

where from Eq. 25,

$$\langle n\mathbf{k}|[\hat{\mathbf{r}}_i, \hat{H}_0]|m\mathbf{k}\rangle = i\delta_{nm}\hbar(\omega_m(\mathbf{k}))_{;\mathbf{k}} = i\delta_{nm}\hbar\nabla_{\mathbf{k}}\omega_m(\mathbf{k}), \tag{B.3}$$

then

$$(\omega_n(\mathbf{k}))_{;\mathbf{k}} = \nabla_{\mathbf{k}}\omega_n(\mathbf{k}). \tag{B.4}$$

Now, from Eq. 19

$$\langle n\mathbf{k}|[\hat{\mathbf{r}}_e, \hat{H}_0]|m\mathbf{k}\rangle = i\hbar\frac{\mathbf{p}_{nm}(\mathbf{k})}{m} \qquad n \neq m, \tag{B.5}$$

and from Eq. 18

$$\langle n\mathbf{k}|[\hat{\mathbf{r}}, \hat{H}_0]|m\mathbf{k}\rangle = i\hbar\frac{\mathbf{p}_{nm}(\mathbf{k})}{m}, \tag{B.6}$$

therefore, substituting above into

$$\langle n\mathbf{k}|[\hat{\mathbf{r}},\hat{H}_0]|m\mathbf{k}\rangle = \langle n\mathbf{k}|[\hat{\mathbf{r}}_i,\hat{H}_0]|m\mathbf{k}\rangle + \langle n\mathbf{k}|[\hat{\mathbf{r}}_e,\hat{H}_0]|m\mathbf{k}\rangle, \tag{B.7}$$

we get

$$i\hbar\frac{\mathbf{p}_{nm}(\mathbf{k})}{m} = i\delta_{nm}\hbar\nabla_{\mathbf{k}}\omega_m(\mathbf{k}) + i\hbar(1-\delta_{nm})\frac{\mathbf{p}_{nm}(\mathbf{k})}{m}, \tag{B.8}$$

from where

$$\frac{\mathbf{p}_{nn}(\mathbf{k})}{m} = \nabla_{\mathbf{k}}\omega_n(\mathbf{k}), \tag{B.9}$$

so from Eq. B.4

$$(\omega_n(\mathbf{k}))_{;k^a} = \frac{p^a_{nn}(\mathbf{k})}{m}. \tag{B.10}$$

## Appendix C.

We obtain the generalized derivative $(\mathbf{r}_{nm}(\mathbf{k}))_{;\mathbf{k}}$. We start with the basic result

$$[r^a,p^b] = i\hbar\delta_{ab}, \tag{C.1}$$

then

$$\langle n\mathbf{k}|[r^a,p^b]|m\mathbf{k}'\rangle = i\hbar\delta_{ab}\delta_{nm}\delta(\mathbf{k}-\mathbf{k}'), \tag{C.2}$$

so

$$\langle n\mathbf{k}|[r^a_i,p^b]|m\mathbf{k}'\rangle + \langle n\mathbf{k}|[r^a_e,p^b]|m\mathbf{k}'\rangle = i\hbar\delta_{ab}\delta_{nm}\delta(\mathbf{k}-\mathbf{k}'). \tag{C.3}$$

From Eq. 25 and 26

$$\langle n\mathbf{k}|[r^a_i,p^b]|m\mathbf{k}'\rangle = i\delta(\mathbf{k}-\mathbf{k}')(p^b_{nm})_{;k^a} \tag{C.4}$$

$$(p^b_{nm})_{;k^a} = \nabla_{k^a}p^b_{nm}(\mathbf{k}) - ip^b_{nm}(\mathbf{k})\left(\xi^a_{nn}(\mathbf{k}) - \xi^a_{mm}(\mathbf{k})\right), \tag{C.5}$$

and

$$\begin{aligned}
\langle n\mathbf{k}|[r^a_e,p^b]|m\mathbf{k}'\rangle &= \sum_{\ell\mathbf{k}''}\Big(\langle n\mathbf{k}|r^a_e|\ell\mathbf{k}''\rangle\langle\ell\mathbf{k}''|p^b|m\mathbf{k}'\rangle \\
&\quad -\langle n\mathbf{k}|p^b|\ell\mathbf{k}''\rangle\langle\ell\mathbf{k}''|r^a_e|m\mathbf{k}'\rangle\Big) \\
&= \sum_{\ell\mathbf{k}''}\Big((1-\delta_{n\ell})\delta(\mathbf{k}-\mathbf{k}'')\xi^a_{n\ell}\delta(\mathbf{k}''-\mathbf{k}')p^b_{\ell m} \\
&\quad -\delta(\mathbf{k}-\mathbf{k}'')p^b_{n\ell}(1-\delta_{\ell m})\delta(\mathbf{k}''-\mathbf{k}')\xi^a_{\ell m}\Big)
\end{aligned}$$

$$= \delta(\mathbf{k}-\mathbf{k}')\sum_{\ell}\Bigg((1-\delta_{n\ell})\xi^a_{n\ell}p^b_{\ell m}$$

$$-(1-\delta_{\ell m})p^b_{n\ell}\xi^a_{\ell m}\Bigg)$$

$$= \delta(\mathbf{k}-\mathbf{k}')\Bigg(\sum_{\ell}\Bigg(\xi^a_{n\ell}p^b_{\ell m} - p^b_{n\ell}\xi^a_{\ell m}\Bigg)$$

$$+p^b_{nm}(\xi^a_{mm}-\xi^a_{nn})\Bigg). \tag{C.6}$$

Using Eqs. C.4 and C.6 into Eq. C.3 gives

$$i\delta(\mathbf{k}-\mathbf{k}')\Bigg((p^b_{nm})_{;k^a} \; -i\sum_{\ell}\Bigg(\xi^a_{n\ell}p^b_{\ell m} - p^b_{n\ell}\xi^a_{\ell m}\Bigg)$$

$$-ip^b_{nm}(\xi^a_{mm}-\xi^a_{nn})\Bigg) = i\hbar\delta_{ab}\delta_{nm}\delta(\mathbf{k}-\mathbf{k}'), \tag{C.7}$$

then

$$(p^b_{nm})_{;k^a} = \hbar\delta_{ab}\delta_{nm} + i\sum_{\ell}\Bigg(\xi^a_{n\ell}p^b_{\ell m} - p^b_{n\ell}\xi^a_{\ell m}\Bigg)$$

$$+ip^b_{nm}(\xi^a_{mm}-\xi^a_{nn}), \tag{C.8}$$

and from Eq. C.5,

$$\nabla_{k^a}p^b_{nm} = \hbar\delta_{ab}\delta_{nm} + i\sum_{\ell}\Bigg(\xi^a_{n\ell}p^b_{\ell m} - p^b_{n\ell}\xi^a_{\ell m}\Bigg). \tag{C.9}$$

Now, there are two cases. We use Eqs. 21 and 22.
*Case $n = m$*

$$\frac{1}{\hbar}\nabla_{k^a}p^b_{nn} = \delta_{ab} - \frac{m}{\hbar}\sum_{\ell}\omega_{\ell n}\Bigg(r^a_{n\ell}r^b_{\ell n} + r^b_{n\ell}r^a_{\ell n}\Bigg), \tag{C.10}$$

that gives the familiar expansion for the inverse effective mass tensor $(m_n^{-1})_{ab}$ [26].
*Case $n \neq m$*

$$(p^b_{nm})_{;k^a} = \hbar\delta_{ab}\delta_{nm} + i\sum_{\ell\neq m\neq n}\Bigg(\xi^a_{n\ell}p^b_{\ell m} - p^b_{n\ell}\xi^a_{\ell m}\Bigg)$$

$$+i\Bigg(\xi^a_{nm}p^b_{mm} - p^b_{nm}\xi^a_{mm}\Bigg)$$

$$+i\Bigg(\xi^a_{nn}p^b_{nm} - p^b_{nn}\xi^a_{nm}\Bigg) + ip^b_{nm}(\xi^a_{mm}-\xi^a_{nn})$$

$$= -m \sum_{\ell} \left( \omega_{\ell m} r^a_{n\ell} r^b_{\ell m} - \omega_{n\ell} r^b_{n\ell} r^a_{\ell m} \right) + i\xi^a_{nm}(p^b_{mm} - p^b_{nn})$$

$$= -m \sum_{\ell} \left( \omega_{\ell m} r^a_{n\ell} r^b_{\ell m} - \omega_{n\ell} r^b_{n\ell} r^a_{\ell m} \right) + imr^a_{nm}\Delta^b_{mn}, \quad (C.11)$$

where

$$\Delta^b_{mn} = \frac{p^b_{mm} - p^b_{nn}}{m}. \quad (C.12)$$

Now, for $n \neq m$, Eqs. 22, B.10 and C.11 and the chain rule, give

$$(r^b_{nm})_{;k^a} = \left(\frac{p^b_{nm}}{im\omega_{nm}}\right)_{;k^a} = \frac{1}{im\omega_{nm}} \left(p^b_{nm}\right)_{;k^a} - \frac{p^b_{nm}}{im\omega^2_{nm}} \left(\omega_{nm}\right)_{;k^a}$$

$$= \frac{i}{\omega_{nm}} \sum_{\ell} \left( \omega_{\ell m} r^a_{n\ell} r^b_{\ell m} - \omega_{n\ell} r^b_{n\ell} r^a_{\ell m} \right) + \frac{r^a_{nm}\Delta^b_{mn}}{\omega_{nm}}$$

$$-\frac{r^b_{nm}}{\omega_{nm}} \left(\omega_{nm}\right)_{;k^a}$$

$$= \frac{i}{\omega_{nm}} \sum_{\ell} \left( \omega_{\ell m} r^a_{n\ell} r^b_{\ell m} - \omega_{n\ell} r^b_{n\ell} r^a_{\ell m} \right) + \frac{r^a_{nm}\Delta^b_{mn}}{\omega_{nm}}$$

$$-\frac{r^b_{nm}}{\omega_{nm}} \frac{p^a_{nn} - p^a_{mm}}{m}$$

$$= \frac{r^a_{nm}\Delta^b_{mn} + r^b_{nm}\Delta^a_{mn}}{\omega_{nm}}$$

$$+\frac{i}{\omega_{nm}} \sum_{\ell} \left( \omega_{\ell m} r^a_{n\ell} r^b_{\ell m} - \omega_{n\ell} r^b_{n\ell} r^a_{\ell m} \right) \quad (C.13)$$

## References


1. For recent reviews see M. Downer, B. S. Mendoza, and V. I. Gavrilenko, Surf. Interface Anal. **31**, 966-986 (2001), and G. Lüpke, Surf. Sci. Reports **35**, 75 (1999), and references therein.
2. B. Mendoza, M. Palummo, R. Del Sole and G. Onida, Phys. Rev. B **63**, 205406-1/6 (2001).
3. M. C. Downer, J. G. Ekerdt, *N. Arzate*, Bernardo S. Mendoza, V. Garvrilenko and R. Wu, Phys. Rev. Lett. **84**, 3406 (2000).
4. V. I. Gavrilenko, R. Q. Wu, M. C. Downer, J. G. Ekerdt, D. Lim, L. Mantese, and P. Parkinson. *Thin Solid Films*, **364**, 1 (2000).
5. B. S. Mendoza, W. L. Mochán, and J. A. Maytorena, Phys. Rev. B **60**, 14334 (1999).
6. B. S. Mendoza, A. Gaggiotti, and R. D. Sole, Phys. Rev. Lett. **81**, 3781 (1998).
7. B. S. Mendoza and W. L. Mochán, Phys. Rev. B **53**, 10473 (1996); ibid **55**, 2489 (1997).

8. P. Guyot-Sionnest, A. Tadjedinne, and A. Liebsch, Phys. Rev. Lett. **64**, 1678 (1990).
9. B. S. Mendoza, Epioptics 2000, Ed. A. Cricenti, World Scientific 2001, ISBN 981-02-4771-0, p. 99-108.
10. N. Arzate and Bernardo S. Mendoza, Phys. Rev. B **63**, 125303-1/14 (2001).
11. C. Aversa and J. E. Sipe, Phys. Rev. B **52**, 14636 (1995).
12. J. E. Sipe and A. I. Shkrebtii, Phys. Rev. B **61**, 5337 (2000).
13. W. R. L. Lambertch and S. N. Rashkeev, phys. stat. sol. (b) **217**, 599 (2000).
14. E. I. Blount, *Solid State Physics: Advances in research and applications* (Academic, New York, 1962) Vol. 13.
15. L. Reining, R. Del Sole, M. Cini, and J. G. Ping, Phys. Rev. B **50**, 8411 (1994).
16. Bernardo S. Mendoza, Maurizia Palummo, Giovanni Onida and Rodolfo Del Sole, Phys. Rev. B **63**, 205406 (2001).
17. J. Mejía, C. Salazar, Bernardo S. Mendoza, Revista Mexicana de Física **50**, 134 (2004).
18. Conor Hogan, Rodolfo Del Sole and Giovanni Onida, Phys. Rev. B **68**, 035405 (2003).
19. C. Castillo, Bernardo S. Mendoza, W. G. Schmidt, P. H. Hahn and F. Bechstedt, Phys. Rev. B **68**, R041310 (2003).
20. Bernardo S. Mendoza, F. Nastos, N. Arzate and J.E. Sipe, Phys. Rev. B **74**, 075318 (2006).
21. J. D. Jackson, *Classical electrodynamics*, John Wiley & Sons, New York, 1975, 2nd Ed. p. 282.
22. S. N. Rashkeev, W. R. L. Lambrecht, and B. Segall, Phys. Rev. B **57**, 3905 (1998).
23. M. Cini, Phys. Rev. B **43**, 4792 (1991).
24. V. Mizrahi and J.E. Sipe J. Opt. Soc. Am. B **5**, 660 (1988).
25. J. Mejía and B.S. Mendoza, Surf. Science **487/1-3**, 180-190 (2001).
26. N.W. Ashcroft and N.D. Mermin, *Solid State Physics*, Saunders College, Philadelphia, 1976.

# PARAMETERFREE CALCULATIONS OF OPTICAL PROPERTIES FOR SYSTEMS WITH MAGNETIC ORDERING OR THREE-DIMENSIONAL CONFINEMENT

F. BECHSTEDT*, C. RÖDL, L.E. RAMOS, F. FUCHS, P.H. HAHN, and J. FURTHMÜLLER

*Institut für Festkörpertheorie und -optik and European Theoretical Spectroscopy Facility (ETSF)*
*Friedrich-Schiller-Universität Jena,*
*Max-Wien-Platz 1, 07743 Jena, Germany*
** E-mail: bech@ifto.physik.uni-jena.de*
*www.ifto.uni-jena.de*

We demonstrate the potential of the electronic-structure methods developed recently for the calculation of optical properties of solids. The many-body effects, i.e. quasiparticles, excitons and local fields, are fully taken into account by solution of the combined Dyson and Bethe-Salpeter equations. The principal effects but also the validity of the used approximations are discussed for bulk semiconductors such as Si and prototypical molecules such as $SiH_4$. The extension of the theory to systems with spin ordering or strong confinement is demonstrated. Their influence on the many-body effects is discussed for the paradigmatic insulator MnO and a silicon nanocrystal.

*Keywords*: many-body perturbation theory, optical properties, magnetic ordering, electron confinement, nanocrystal

## 1. Introduction

Recent years have seen impressive methodological progress in the accurate numerical modelling of optical properties from *first principles* using many-body perturbation theory (MBPT) [1]. It has become possible to compute single-particle electronic excitations in an accurate manner using Hedin's GW approximation (GWA) [2]. In addition, the Bethe-Salpeter equation (BSE) for electron-hole pair excitations can be solved in the framework of the same approximation, in order to account for excitonic and local-field (LF) contributions to the polarization function [3–5]. However, the large numerical effort required to solve the BSE has restricted such calculations to the interaction of relatively few electron-hole pairs. Therefore, the calculations of optical properties are usually limited to bulk semiconductors [6–13] or to semiconductor surfaces with strongly localized and energetically well separated states [14,15]. There are only first trials to include more pair states in the calculation of optical spectra, for instance to describe surface-modified bulk excitons [16,17]. The formalism, however, also works for strongly localized electron-hole pair excitations in highly polarizable solids [18] and molecules [19].

In this paper the progress in the calculation of optical properties is discussed with a focus on novel materials with magnetic ordering and nanostructures. This concerns the determination of the geometries and electronic structures but also the inclusion of many-body effects in the framework of MBPT. We show that optical spectra such as the optical absorption can now be calculated from *first principles.* One important intermediate step is related to the quasiparticle band structures. Besides the semiconductor silicon the antiferromagnet MnO and the nanocrystal $Si_{17}H_{36}$ are considered as examples. In Sections 2 and 3 the computational and modelling methods are described. We discuss the validity of approximations applied usually. The resulting quasiparticle band structures and optical absorption spectra are discussed in detail in Sections 4 and 5. A brief summary concludes the paper in Section 6.

## 2. Theoretical and computational methods

In order to calculate the optical spectra we proceed in three steps. In the first step we derive the optimized atomic geometry of the system and a starting electronic structure in Kohn-Sham approximation [20]. We take advantage of the translational symmetry. The Bloch picture is valid with wave vectors $\mathbf{k}$ from the Brillouin zone (BZ) and band indices $\nu$ as good quantum numbers. Within the local density approximation (LDA) to exchange and correlation the single-particle Kohn-Sham (KS) equation yields eigenfunctions $|\nu\mathbf{k}\rangle$ (or $\psi_{\nu\mathbf{k}}(\mathbf{x}) = \langle\mathbf{x}|\nu\mathbf{k}\rangle$) and eigenvalues $\varepsilon_\nu(\mathbf{k})$. In the next step we account for the excitation aspect. The electron-electron interaction leads to quasiparticles. Finally, the mutual interactions of the excited quasielectrons and quasiholes are also treated.

### 2.1. *Ground state*

First, the energetically favored geometry is determined. The required total-energy and electronic-structure calculations are based on density functional theory (DFT) [21] and the LDA [20]. The implementation of DFT is based on the Vienna Ab-initio Simulation Package (VASP) [22]. The pseudopotentials are generated according to the projector-augmented wave (PAW) method to describe the interaction between atomic cores and valence electrons [23]. All-electron PAW wave functions are used to determine the optical matrix elements [24]. The energy cutoffs of the plane-wave expansions of the Bloch states to obtain converged results are relatively low.

### 2.2. *Quasiparticles*

The optically excited electrons and holes also interact as individual particles with the inhomogeneous electron gas of the system. This leads to an exchange-correlation self-energy $\Sigma(\varepsilon)$ of the system beyond the exchange-correlation potential $V_{\mathrm{XC}}$ that is already included in the Kohn-Sham equation of the DFT-LDA. The corresponding renormalization gives rise to quasielectrons and quasiholes [1,2]. Assuming the same energetic ordering of the quasiparticle and the KS energies, the quasiparticle effects are included within first-order perturbation theory [25,26]. The KS wave functions are not updated [27] and the DFT-LDA eigenvalues are corrected according to

$$\varepsilon_{\nu}^{QP}(\mathbf{k}) = \varepsilon_{\nu}(\mathbf{k}) + \Delta_{\nu}(\mathbf{k}), \tag{1}$$

$$\Delta_{\nu}(\mathbf{k}) = \mathrm{Re}\langle \nu\mathbf{k}|\Sigma\left(\varepsilon_{\nu}^{QP}(\mathbf{k})\right) - V_{\mathrm{XC}}|\nu\mathbf{k}\rangle. \tag{2}$$

The exchange-correlation self-energy operator $\Sigma$ is taken within the GWA [2]. In the explicit calculations we introduce further approximations following the schemes by Hybertsen and Louie [28] and Bechstedt et al. [29] or Cappellini et al. [30]. For several semiconductors and their surfaces this approximate treatment of the self-energy corrections (2) has been shown to result in excitation energies which are within 0.1 eV of the experimental values [13,31]. The imaginary part of the self-energy matrix elements $\mathrm{Im}\langle \nu\mathbf{k}|\Sigma(\varepsilon)|\nu\mathbf{k}\rangle$ can give rise to a finite lifetime of the quasiparticle.

### 2.3. *Electron-hole pairs and dielectric function*

The frequency-dependent macroscopic dielectric function is related to the polarization function $P$. Within the Bloch picture one obtains for a crystalline system [5,17]

$$\begin{aligned}\varepsilon_{\alpha\beta}(\omega) = \delta_{\alpha\beta} - \frac{8\pi e^2\hbar^2}{V}\sum_{c,v,\mathbf{k}}\sum_{c',v',\mathbf{k}'}\Big\{ & M_{cv}^{\alpha^*}(\mathbf{k})M_{c'v'}^{\beta}(\mathbf{k}')P(cv\mathbf{k},c'v'\mathbf{k}';\omega) \\ & + c.c. \quad \text{and} \quad \omega \leftrightarrow -\omega\Big\}\end{aligned} \tag{3}$$

with matrix elements of the velocity operator $\boldsymbol{v}$

$$M_{cv}^{\alpha}(\mathbf{k}) = \frac{\langle c\mathbf{k}|v_{\alpha}|v\mathbf{k}\rangle}{\varepsilon_c(\mathbf{k}) - \varepsilon_v(\mathbf{k})} \tag{4}$$

and $V$ as the normalization volume. In (3) the sums run over pairs of electrons in empty conduction band states $|c\mathbf{k}\rangle$ and holes in occupied valence band states $|v\mathbf{k}\rangle$, which are virtually or physically excited by photons. The effect of the photon wave vector is neglected.

The polarization function $P$ obeys a BSE. Since the self-energy $\Sigma(\varepsilon) \sim GW$ its kernel is dominated by the single-particle Green's function $G$ and the dynamically screened Coulomb potential $W$, by $\delta\Sigma/\delta G \sim W + G\delta W/\delta G$. Usually the last term is neglected [5,32] and the screening in $W$ is replaced by the static one [5,33,34]. Neglecting in addition the coupling of resonant and antiresonant electron-hole pairs as well as the non-particle-conserving contributions to the electron-hole interaction [6], the BSE reads as

$$\sum_{c'',v'',\mathbf{k}''} \{H(cv\mathbf{k}, c''v''\mathbf{k}'') - \hbar(\omega + i\gamma)\delta_{cc''}\delta_{vv''}\delta_{\mathbf{k}\mathbf{k}''}\} P(c''v''\mathbf{k}'', c'v'\mathbf{k}'; \omega) = -\delta_{cc'}\delta_{vv'}\delta_{\mathbf{k}\mathbf{k}'} \tag{5}$$

with the effective electron-hole pair Hamiltonian $H(cv\mathbf{k}, c'v'\mathbf{k}')$ and a small damping $\gamma$ of the pair excitations. The Hamiltonian of pairs of excited electrons and holes, more precisely of quasielectrons and quasiholes, is given by [3,6,19]

$$\begin{aligned} H(cv\mathbf{k}, c'v'\mathbf{k}') = {} & \left[\varepsilon_c^{QP}(\mathbf{k}) - \varepsilon_v^{QP}(\mathbf{k})\right] \delta_{cc'}\delta_{vv'}\delta_{\mathbf{k}\mathbf{k}'} \\ & -W(cv\mathbf{k}, c'v'\mathbf{k}') + 2\bar{v}(cv\mathbf{k}, c'v'\mathbf{k}') \end{aligned} \tag{6}$$

with the matrix elements

$$W(cv\mathbf{k}, c'v'\mathbf{k}') = \int d^3\mathbf{x} \int d^3\mathbf{x}' \psi_{c\mathbf{k}}^*(\mathbf{x})\psi_{c'\mathbf{k}'}(\mathbf{x})W(\mathbf{x}, \mathbf{x}')\psi_{v\mathbf{k}}(\mathbf{x}')\psi_{v'\mathbf{k}'}^*(\mathbf{x}') \tag{7}$$

and

$$\bar{v}(cv\mathbf{k}, c'v'\mathbf{k}') = \int d^3\mathbf{x} \int d^3\mathbf{x}' \psi_{c\mathbf{k}}^*(\mathbf{x})\psi_{v\mathbf{k}}(\mathbf{x})\bar{v}(\mathbf{x} - \mathbf{x}')\psi_{c'\mathbf{k}'}(\mathbf{x}')\psi_{v'\mathbf{k}'}^*(\mathbf{x}') \tag{8}$$

of the (statically) screened Coulomb interaction $W(\mathbf{x}, \mathbf{x}')$ [35] and a bare Coulomb interaction $\bar{v}(\mathbf{x} - \mathbf{x}')$. Only the short-range part of the latter is taken into account in agreement with the physical character of expression (8) as electron-hole exchange [4].

The screened contribution $W$ (7) to the total electron-hole interaction includes the classical attraction of electron and hole as represented by the

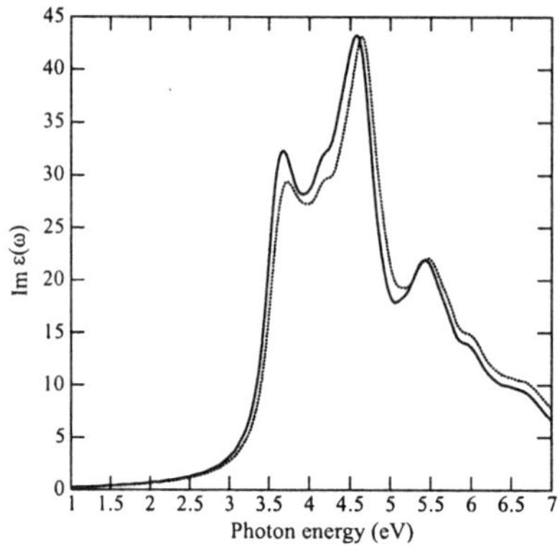


Fig. 1. Imaginary part of the dielectric function of a Si crystal. Solid line: usual BSE (5) with kernel $\sim W$, dotted line: the term $G\delta W/\delta G$ is also included in the kernel.

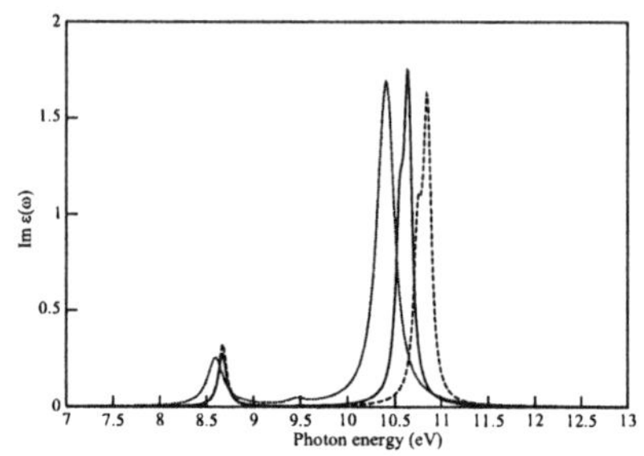


Fig. 2. Imaginary part of the dielectric function for silane $SiH_4$. Solid line: usual BSE (5), dotted line: with dynamical screening in Shindo approximation [33,34], dashed line: with coupling between resonant and antiresonant contributions to the polarization function.

diagonal elements $c = c'$ and $v = v'$. It is responsible for the electron-hole binding in the Wannier-Mott limit [36]. The other contributions represent the mixing of electron-hole pairs which is responsible for the redistribution of oscillator strength in optical spectra [6,7,17]. The electron-hole exchange term $\propto \bar{v}$ (8) [3,37,38] corresponds to the inclusion of LF effects [3,39, 40]. Indeed, in the bulk case it has been shown [38] that the inclusion of $\bar{v}$ in the BSE (5) gives a polarization corresponding to the macroscopic dielectric susceptibility. Effects that have been neglected in the BSE (5) and the two-particle Hamiltonian (6) are shown in Figs. 1 and 2 for bulk Si and the $SiH_4$ molecule, respectively. The figures show that the influence of $G\delta W/\delta G$ (Fig. 1) is negligible for crystals while other effects such as dynamical screening and the coupling of resonant and antiresonant terms may compensate each other for small molecules (Fig. 2).

An extremely important problem in modelling excitonic interactions according to (7) and (8) is to reach convergence with respect to the number of pair states $cv\mathbf{k}$ taken into account. Usually one expects converged results for the optical spectra for a mesh of $\mathbf{k}$ points in the Brillouin zone which gives a good interband density of states. Our experience shows that this criterion is insufficient in the presence of many-body effects. One has to take into account at least 1000 mesh points [13]. For systems whose imaginary part of the frequency-dependent dielectric function is nearly constant in certain frequency regions, one has to increase the number of mesh points by one order of magnitude [11].

## 3. Modelling of spin and confinement

### 3.1. *Spin ordering*

We consider the situation of a spin-polarized system. In the limit of collinear spins the formulas and equations of the spin-less case described above can be widely taken over. Only the single-particle wave functions ($\psi_{\nu\mathbf{k}}(\mathbf{x}) \rightarrow \psi_{\nu\mathbf{k}m_s}(\mathbf{x})$) and eigenvalues ($\varepsilon_\nu(\mathbf{k}), \varepsilon_\nu^{QP}(\mathbf{k}) \rightarrow \varepsilon_{\nu m_s}(\mathbf{k}), \varepsilon_{\nu m_s}^{QP}(\mathbf{k})$) depend additionally on the spin quantum number $m_s$. However, all matrix elements of the velocity operator remain diagonal in the spin index. In the collinear case no spin umklapp processes occur. The two-particle Hamiltonian takes the form

$$\begin{aligned} H(cv\mathbf{k}m_s, c'v'\mathbf{k}'m_s') = & \left[\varepsilon_{cm_s}^{QP}(\mathbf{k}) - \varepsilon_{vm_s}^{QP}(\mathbf{k})\right]\delta_{cc'}\delta_{vv'}\delta_{\mathbf{k}\mathbf{k}'}\delta_{m_s m_s'} \\ & - W(cv\mathbf{k}m_s, c'v'\mathbf{k}'m_s)\delta_{m_s m_s'} \\ & + \bar{v}(cv\mathbf{k}m_s, c'v'\mathbf{k}'m_s') \end{aligned} \tag{9}$$

The spin dependence of the electron-hole exchange interaction (i.e., local-field effects) in (9) does not allow anymore a diagonalization of the Hamiltonian resulting in uncoupled singlet and triplet pair states. However, in the collinear case the effect on optical properties is less important since the single-particle spin is not changed during the optical excitation of an electron. The solutions of the eigenvalue problem of (9)

$$\sum_{c',v',\mathbf{k}',m_s'} H(cv\mathbf{k}m_s, c'v'\mathbf{k}'m_s')A_\Lambda(c'v'\mathbf{k}'m_s') = E_\Lambda A_\Lambda(cv\mathbf{k}m_s) \tag{10}$$

can be used to reformulate the dielectric function as in the spin-less case. One finds for the diagonal elements [41]

$$\begin{aligned} \varepsilon_{\alpha\alpha}(\omega) = 1 + \frac{4\pi e^2\hbar^2}{V}\sum_\Lambda \left|\sum_{c,v,\mathbf{k},m_s} M_{cvm_s}^{\alpha^*}(\mathbf{k})A_\Lambda(cv\mathbf{k}m_s)\right|^2 \times \\ \times \left[\frac{1}{E_\Lambda - \hbar(\omega + i\gamma)} + \frac{1}{E_\Lambda + \hbar(\omega + i\gamma)}\right]. \end{aligned} \tag{11}$$

Each excitonic excitation is weighted by the sum over the two possible spin orientations of the electron excited from the valence into the conduction band.

### 3.2. *Nanocrystals*

The plane-wave expansion of the electronic eigenfunctions requires a periodic arrangement of supercells containing the nanocrystal that should be modelled as shown in Fig. 3. Their edge lengths have to be chosen large enough to avoid the interaction of such a nanoobject with its image in neighboring supercells. We typically use a distance of about 1 nm between the surfaces of neighboring nanocrystals. This avoids a broadening of levels into bands. The challenge is to treat numerically the electronic-structure problems for the resulting large supercells. Unfortunately, despite the large cells the scattering states of nanocrystals are not yet correctly described in the supercell approximation. For that reason we restrict our studies to small photon energies which do not give rise to a photoionization of the nanoobject. The advantage of dealing only with dispersionless levels is that all treatments of many-body interactions can be restricted to the center of the Brillouin zone.

The nanocrystals are constructed starting from a central Si atom and adding its nearest neighbors, thereby assuming tetrahedral bonding. This procedure is repeated atomic shell by atomic shell up to the outermost one, where the dangling bonds are saturated by hydrogen atoms. The assumption of a local diamond-like structure is in accordance with recent transmission electron microscopy studies [42,43]. The construction procedure leads to faceted clusters $SiH_4$, $Si_5H_{12}$, $Si_{17}H_{36}$, $Si_{41}H_{60}$, $Si_{83}H_{108}$, ... .

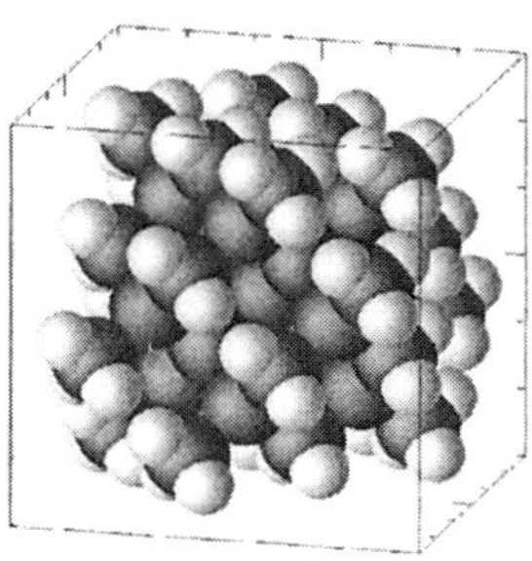

Fig. 3. Si nanocrystal passivated with H atoms in a cubic supercell.

## 4. Example: Antiferromagnetic insulator MnO

The high-temperature (above Néel temperature of 118 K) phase of MnO is a paramagnetic system. The minimization of the total energy indicates a type-II antiferromagnetic insulator with a slightly rhombohedrally distorted rocksalt (B1) structure as the ground state (i.e., low-temperature phase). The magnetic ordering is a result of parallel spins in (111) planes as shown in Fig. 4. The rhombohedral distortion takes the form of a compression along the [111] direction with a deviation of 1.8° from the cubic angles. The theoretical lattice constant is about 8.87 Å and the local magnetic moment has a maximum value of 4.4 $\mu_B$ in good agreement with other first principles calculations and experimental data (see e.g. [44] and references therein). The Kohn-Sham band structure indeed describes an insulator with an indirect $T-\Gamma$ gap of 0.7 eV and direct gaps of 1.2 ($L-L$) or 1.4 ($\Gamma-\Gamma$) eV. The calculated quasiparticle band structure is shown in Fig. 5. The gaps are opened by about 2.5 eV to 3.1, 4.5, and 3.9 eV, respectively. Experimentally slightly larger fundamental gaps are observed in several spectroscopies.

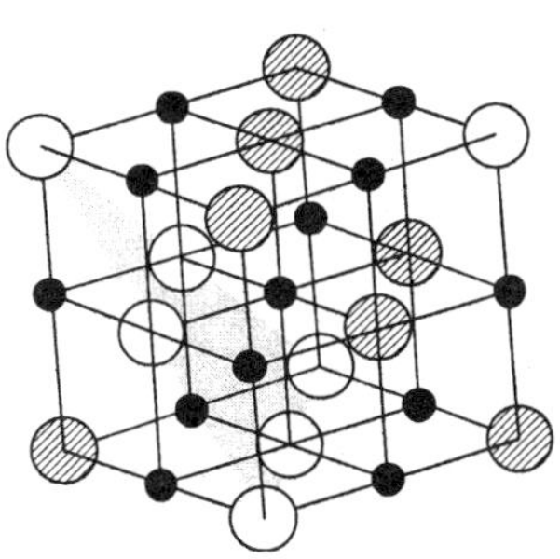

Fig. 4. Antiferromagnetic type-II ordering of the Mn atoms (large open and shaded circles) with parallel spins within (111) planes of MnO.

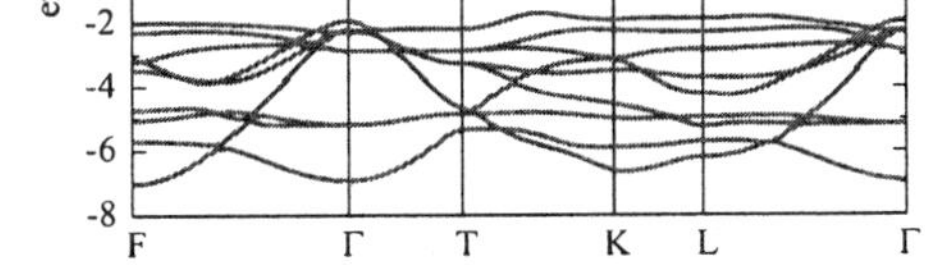


Fig. 5. Quasiparticle band structure of the antiferromagnetic insulator MnO.

The calculated optical absorption spectra, more precisely imaginary parts of the averaged dielectric function, are shown in Fig. 6. The inclusion of quasiparticle effects shifts the entire spectrum more or less by about 2.5 eV towards higher energies. The Coulomb correlation of the quasielectron-quasihole pairs leads almost to an overall redshift. In addition, spectral strength is redistributed from higher photon energies to lower energies. Basically we make the same observations as in the case of other semiconductors and insulators independent of their bonding character and their crystal structure [12,13]. The comparison with experimental spectra [45,46] indicates more or less the same peak positions but an overestimation of spectral strength. However, bound states are not observed. The reasons are that we study only electric-dipole-allowed optical transitions. Intraatomic $d-d$ transitions do not occur within our approximation of optical properties. The same holds for spin-umklapp processes induced by spin-orbit interaction.

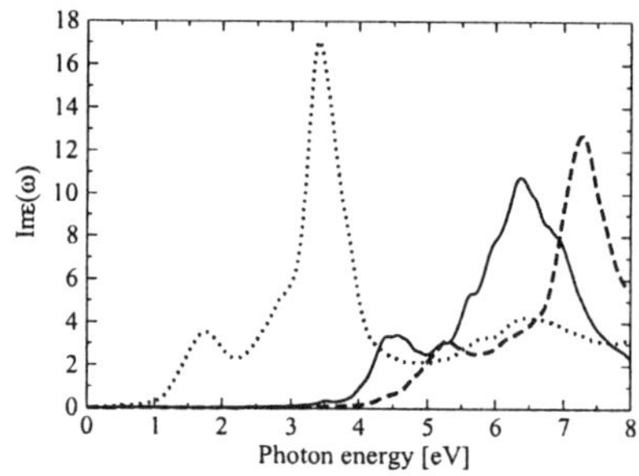


Fig. 6. Imaginary part of the dielectric function of MnO in three different approximations. Dotted line: independent-particle approximation, dashed line: independent-quasiparticle approximation (including QP shifts), and solid line: Coulomb-correlated electron-hole pairs with electron-hole attraction and electron-hole exchange.

## 5. Example: Nanocrystal $Si_{17}H_{36}$

The excitation of electrons or holes leads to drastic renormalization of their energies. This is clearly demonstrated for the real part of the self-energy in Fig. 7. The hole states exhibit a much larger frequency dispersion and shifts towards lower energies in comparison to electron states. As a consequence the hole states have more pronounced satellite structures while the electron states possess a quasiparticle renormalization factor [25,26] of about 1. The so-called Hartree-Fock energy $\varepsilon_\lambda^{\mathrm{HF}}$ contains only a shift of type (2) where the correlation part of the self-energy is neglected. The crossings of the straight line $\varepsilon + (\varepsilon_\lambda^{\mathrm{HF}} - \varepsilon_\lambda)$ with the self-energy curves in Fig. 7 yield the quasiparticle energies $\varepsilon_\lambda^{\mathrm{QP}}$. In the energy range of the quasiparticle energies the imaginary part of the self-energy can be neglected [19].

The resulting optical absorption spectrum is shown in Fig. 8. At a first glance one observes similar effects as for the bulk crystals (cf. Fig. 6). The quasiparticle effects induce a remarkable blueshift of the spectrum while the electron-hole pair interactions lead to an overall redshift. However, in addition the Coulomb interactions between electrons and holes give rise to a considerable broadening of the spectrum accompanied by a more complicated fine-structure. The reason for these findings is the interplay between the screened Coulomb attraction (7) and the repulsive electron-hole exchange (8). Qualitatively the optical absorption with the Coulomb effects looks similar to the spectra obtained within the time-dependent density functional theory in local density approximation (TDLDA) [47,48]. This indicates the much more important role of electron-hole exchange and, hence, local-field effects in the case of nanocrystals.

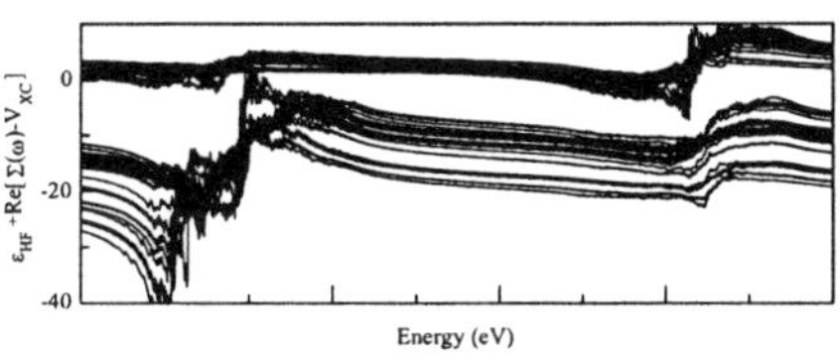


Fig. 7. Energy dependence of the real part of the self-energy difference $\Sigma(\varepsilon) - V_{XC}$ for the single-particle states $\lambda$ in $Si_{17}H_{36}$.

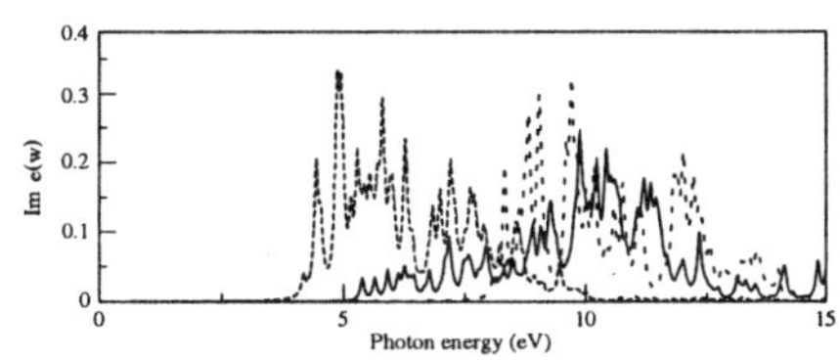


Fig. 8. Imaginary part of the dielectric function of $Si_{17}H_{36}$ in three different approximations. Dotted line: independent-particle approximation, dashed line: independent-quasiparticle approximation (including QP shifts), and solid line: Coulomb-correlated electron-hole pairs with electron-hole attraction and electron-hole exchange.

## 6. Summary

We have demonstrated recent progress in the theoretical description of optical properties of crystalline or nanocrystalline materials using parameter-free computations under inclusion of the necessary many-body effects. Spin polarization has been taken into account on the level of collinear spins. Nanoobjects have been modelled using supercells. The calculations are based on density functional theory in a certain approximation for exchange and correlation. They yield the lattice constants (in general, the atomic positions) and the Kohn-Sham band structures which neglect the excitation aspect. After optical excitation of electron-hole pairs each excited particle interacts with the remaining electron system. This leads to a renormalization to a quasiparticle, a quasielectron or quasihole. Its main effect is a shift of the Kohn-Sham eigenvalues towards quasiparticle energies, and, hence, an opening of all gaps and a blueshift of transition energies. The quasi-electrons and quasiholes interact via two different mechanisms. There is a direct screened Coulomb attraction that yields an overall redshift of the absorption spectra with respect to curves for independent quasiparticles. Oscillator/spectral strength is redistributed from higher photon energies to lower ones. In addition, there also acts an electron-hole exchange that can be identified with the modelling of the local-field effects in the corresponding two-particle Hamiltonian. In any case, their inclusion allows to compute the macroscopic dielectric function instead of the microscopic one. Despite the restriction to collinear spins the electron-hole exchange couples different single-particle spins. However, its effect on the absorption spectrum remains small. This is in contrast to the case of nanocrystals where this interaction contributes to a redistribution of spectral strength over a wider range of photon energies.

## Acknowledgement

The authors acknowledge financial support of the EU NoE NANOQUANTA (contract No. NM4-CT-2004-500198) and a grant of computer time at NIC Jülich (project HJN21/2005).

## References

1. G. Onida, L. Reining, and A. Rubio, *Rev. Mod. Phys.* **74**, 601 (2002)
2. L. Hedin, *Phys. Rev.* **139**, A796 (1965); L. Hedin and S. Lundqvist, *Solid State Phys.* **23**, 1 (1969).
3. L.J. Sham and T.M. Rice, *Phys. Rev.* **144**, 708 (1966).
4. W. Hanke and L.J. Sham, *Phys. Rev. B* **21**, 4656 (1980).
5. G. Strinati, *Riv. Nuovo Cimento* **11**, 1 (1988).
6. S. Albrecht, L. Reining, R. Del Sole, and G. Onida, *Phys. Rev. Lett.* **80**, 4510 (1998).
7. L.X. Benedict, E.L. Shirley, and R.B. Bohn, *Phys. Rev. Lett.* **80**, 4514 (1998); *Phys. Rev. B* **57**, R9385 (1998).
8. M. Rohlfing and S.G. Louie, *Phys. Rev. Lett.* **81**, 2312 (1998).
9. B. Arnaud and M. Alouani, *Phys. Rev. B* **63**, 085208 (2001).
10. P. Puschnig and C. Ambrosch-Draxl, *Phys. Rev. B* **66**, 165105 (2002).
11. J. Furthmüller, P.H. Hahn, F. Fuchs, and F. Bechstedt, *Phys. Rev. B* **72**, 205106 (2005).
12. F. Bechstedt, K. Seino, P.H. Hahn, and W.G. Schmidt, *Phys. Rev. B* **72**, 245114 (2005).
13. P.H. Hahn, K. Seino, W.G. Schmidt, J. Furthmüller, and F. Bechstedt, *phys. stat. sol. (b)* **242**, 2720 (2005).
14. M. Rohlfing and S.G. Louie, *Phys. Rev. Lett.* **83**, 856 (1999).

15. M. Rohlfing, M. Palummo, G. Onida, and R. Del Sole, *Phys. Rev. Lett.* **85**, 5440 (2000).
16. P.H. Hahn, W.G. Schmidt, and F. Bechstedt, *Phys. Rev. Lett.* **88**, 016402 (2002).
17. W.G. Schmidt, S. Glutsch, P.H. Hahn, and F. Bechstedt, *Phys. Rev. B* **67**, 085307 (2003).
18. P.H. Hahn, W.G. Schmidt, K. Seino, M. Preuss, F. Bechstedt, and J. Bernholc, *Phys. Rev. Lett.* **94**, 037404 (2005).
19. P.H. Hahn, W.G. Schmidt, and F. Bechstedt, *Phys. Rev. B* **72**, 245425 (2005).
20. W. Kohn and L.J. Sham, *Phys. Rev.* **140**, A1133 (1965).
21. P. Hohenberg and W. Kohn, *Phys. Rev.* **136**, B864 (1964).
22. G. Kresse and J. Furthmüller, *Comput. Mater. Sci.* **6**, 15 (1996); *Phys. Rev. B* **54**, 11169 (1996).
23. G. Kresse and D. Joubert, *Phys. Rev. B* **59**, 1758 (1999).
24. B. Adolph, J. Furthmüller, and F. Bechstedt, *Phys. Rev. B* **63**, 125108 (2001).
25. M.S. Hybertsen and S.G. Louie, *Phys. Rev. B* **34**, 5390 (1986)
26. F. Bechstedt, Festkörperprobleme, *Adv. Solid State Phys.*, Vol. **32** (Vieweg, Braunschweig, 1992), p. 161.
27. O. Pulci, F. Bechstedt, G. Onida, R. Del Sole, and L. Reining, *Phys. Rev. B* **60**, 16758 (1999).
28. M.S. Hybertsen and S.G. Louie, *Phys. Rev. B* **37**, 2733 (1988).
29. F. Bechstedt, R. Del Sole, G. Cappellini, and L. Reining, *Solid State Commun.* **84**, 765 (1992).
30. G. Cappellini, R. Del Sole, L. Reining, and F. Bechstedt, *Phys. Rev. B* **47**, 9892 (1993).
31. W.G. Schmidt, J.L. Fattebert, J. Bernholc, and F. Bechstedt, *Surf. Rev. Lett.* **6**, 1159 (1999).
32. G. Baym and L.P. Kadanoff, *Phys. Rev.* **124**, 287 (1961).
33. K. Shindo, *J. Phys. Soc. Japan* **29**, 287 (1970).
34. R. Zimmermann, *phys. stat. sol. (b)* **48**, 603 (1971).
35. F. Bechstedt, K. Tenelsen, B. Adolph, and R. Del Sole, *Phys. Rev. Lett.* **78**, 1528 (1997).
36. G.D. Mahan, *Many-Particle Physics* (Plenum Press, New York, 1990).
37. W. Hanke and L.J. Sham, *Phys. Rev. B* **12**, 4501 (1975).
38. R. Del Sole and E. Fiorino, *Phys. Rev. B* **29**, 4631 (1984).
39. S.L. Adler, *Phys. Rev.* **126**, 413 (1962).
40. N. Wiser, *Phys. Rev.* **129**, 62 (1963).
41. C. Rödl, *Diploma thesis* (Friedrich-Schiller-Universität Jena, 2005).
42. H. Hofmeister and P. Ködderitzsch, *Nanostructured Materials* **12**, 203 (1999).
43. K.C. Scheer, R.A. Rao, R. Muralidhar, S. Bagchi, J. Conner, L. Lozano, C. Perez, M. Sadd, and B.E. White Jr., *J. Appl. Phys.* **93**, 5637 (2003).
44. J.E. Pask, D.J. Singh, I.I. Mazin, C.S. Hellberg, and J. Kortus, *Phys. Rev. B* **64**, 024403 (2001).
45. Ya.M. Ksendzov, I.L. Korobova, K.K. Sidorin, G.P. Startsev, *Fiz. Tverd. Tela* (Leningrad) **18**, 173 (1976).
46. L. Messick, W.C. Walker, and R. Glosser, *Phys. Rev. B* **6**, 3941 (1972).

47. I. Vasiliev, S. Öğüt, and J.R. Chelikowsky, *Phys. Rev. Lett.* **86**, 1813 (2001).
48. L.X. Benedict, A. Puzder, A.J. Williamson, J.C. Grossman, G. Galli, J.E. Klepeis, J.-Y. Raty, and O. Pankratov, *Phys. Rev. B* **68**, 085310 (2003).

# EXCITED STATE PROPERTIES CALCULATIONS: FROM 0 TO 3 DIMENSIONAL SYSTEMS

M. MARSILI [(1)], V. GARBUIO [(1)], M. BRUNO [(1)], O. PULCI [(1)], M. PALUMMO [(1)], E. DEGOLI [(2)], E. LUPPI [(3)], R. DEL SOLE [(1)]

[(1)] *ETSF and INFM and Dipartimento di Fisica dell'Università di Roma Tor Vergata, via della Ricerca Scientifica, I-00133 Roma, Italy*

[(2)] *INFM-S[3] "nanoStructures and bioSystems at Surfaces", Dipartimento di Scienze e Metodi dell'Ingegneria, via G. Amendola 2, Universitá di Modena e Reggio Emilia*

[(3)] *ETSF and Laboratoire des Solides Irradiés, École Polytechnique, F-91128 Palaiseau, France*

We review the theoretical framework of *ab-initio* excited state properties calculations showing the application of these methods to systems of different dimensionality. Density functional theory within the usual and generalised Kohn-Sham scheme is presented and applied to the case of the cleavage surface of diamond. Many body perturbation theory within the GW approximation of the self energy for the calculation of electronic band structures is presented and applied to liquid water. Finally, the Bethe Salpeter equation for the computation of optical properties is presented and used for the case of silicon nanowires and silicon nanoclusters

## 1. Introduction

Many experimental techniques in solid state physics, such as angle resolved photoemission spectroscopy (ARPES), electron energy loss (EELS), optical absorption etc... probe the electronic excitations of the systems under investigation. As a consequence a material is often characterised by its excited state properties. A completely *ab-initio*, parameter free, determination of the excited state properties of materials is thus very important for the interpretation of the experimental spectra and/or for the prediction of the material's features. In this paper we will review the well established theoretical background of the *state of the art* calculation of excited state properties and we will show how these methods can be applied to systems of different dimensionality.

All the calculations presented in this paper are performed using codes[1]

that employ plane wave basis set. This kind of basis set allows for the intensive use of fast algorithms such as fast Fourier transforms (FFT) and moreover allows a systematic check on the convergency of the calculation. Of course plane waves are suited to describe crystal bulk systems in which a certain unit cell is repeated in all the three directions in space. However as we will show below, also lower dimensionality systems such as surfaces, nanowires, and molecules and even disordered systems such as liquid water can be studied within the same scheme. Of course in all these cases special care has to be put in order to avoid spurious effects due to the infinite reproduction of the unit cell, the so called *supercell*. In the case of a surface, the supercell is made by a certain amount of atomic layers meant to reproduce the bulk and the surface of the system and by a certain amount of vacuum. When this supercell is repeated in all the three directions in space the resulting system is given by a infinite series of repeated slabs. To obtain well converged results the vacuum must be deep enough so that the slabs are decoupled, and there must be enough atomic layers so that the bulk is well represented. Also in the case of nanowires, molecules and nanoclusters the final geometry consists of an infinite series of repeated replicas of the system. Like in the case of surfaces, the convergency with respect to the amount of vacuum present between them must be carefully checked. Especially in the calculations of excited state properties, where the electronic empty states play an important role, it might be very convenient to use spherical[4](for molecules and nanoclusters) or cylindrical[5,6] (for nanowires) cutoff functions for treating the long range term of the Coulomb interaction. The peculiar problem related to the study of a liquid disordered system relies on the fact that one should in principle use a very large unit cell. A possible solution to avoid such huge unit cells has been proven[7] to be the use of smaller unit cells but exploiting several molecular dynamics (MD) snapshots of water as input geometries and averaging the results over all the configurations.

The paper is organised as follows: first, ground state calculations within density functional theory (DFT), preliminary to all the excited state calculations will be presented; as an example we will show the calculations for the (111)2×1 surface of diamond. Then, we will see how quasiparticle effects can be introduced to get a reliable description of the electronic band structure and we will see how this method is applied to the case of liquid water. Finally we will see how many body effects, such as the excitonic effects, can be introduced in the calculations of optical spectra, and we will show the results of calculations for silicon nanowires and a small silicon nanocluster.

## 2. Determination of the ground state: C(111)2 × 1

As already mentioned, the computation of the excited state properties is based on a previous well converged ground state calculation. Here we schematically review the theory behind this kind of calculations and show the application for the cleavage surface of diamond.

Within the Born-Oppenheimer approximation the electronic Hamiltonian of a system of interacting electrons in an external potential is given by

$$H = \sum_i -\frac{1}{2}\nabla^2_{\mathbf{r}_i} + V_{ext} + \frac{1}{2}\sum_{i,j}{}' \frac{1}{|\mathbf{r}_i - \mathbf{r}_j|} = T + V_{ext} + V_{e-e}. \qquad (1)$$

Even though the ions coordinates appear just as parameters in $V_{ext}$, eq. (1) is still very involved and a direct diagonalisation to find its eigenstates is still a formidable task that for complex systems is out of the reach of the computational power. The density functional approach, focusing the attention on the *simpler* total electronic density $n(r)$ rather than on the more complex many body wavefunction $\Psi(r_1, ..., r_N)$, opens the way to the study of the ground state properties of even very complex systems.
Density Functional Theory (DFT) treats the case of an external, time independent potential, and is based on the seminal paper of Hohenberg and Kohn of 1964[8]. For a review on DFT see for example[9]. The Hohenberg and Kohn theorem states that once the mutual interaction among the electrons is fixed, in a non time dependent situation and assuming that all the potentials acting on the system are local, all the ground state properties of an interacting electronic system, including, in principle, the many-body wave function, could be expressed as unique functionals of the electronic density alone. In particular this assertion is valid also for the total energy $E$ of the system. For the total energy functional $E[n]$, a variational principle could be derived stating that for a given density $n(\mathbf{r})$

$$E[n(\mathbf{r})] = F[n] + \int d\mathbf{r} V_{ext}(\mathbf{r})n(\mathbf{r}) \geq E_{GS} \qquad (2)$$

where $E_{GS}$ is the ground state energy of the system, and $F[n] = T + V_{ee}$ is a universal functional that contains the contribution of the kinetic ($T$) and electron-electron ($V_{ee}$) part of the hamiltonian. In particular the energy functional finds its minimum at the ground state density $E[n_{GS}] = E_{GS}$. This theorem leads, in principle, to a straightforward method for computing ground state properties: given an approximation for the functional dependence of $F$, the ground state density can be obtained by a minimisation procedure of the energy functional, and then, knowing the functional dependence on the density of the properties of interest, one can find their ground state value. However finding a suitable approximation for $F$ is a formidable and very delicate task. In fact, beside the contribution of the electron-electron interaction, most of which usually can be included in the Hartree term, $F$ contains the kinetic energy whose functional dependence on the density is unknown and, most important, it constitutes a non negligible part of the total energy. This means that an approximation made directly on this term can lead to big errors in the total energy of the system.

The breakthrough appeared one year later, with the paper of Kohn and Sham[10] (KS) where they introduced a different separation of the terms contributing to the total energy $E$. The two main advantages of this separation were first of all to provide a single particle (selfconsistent) scheme to obtain the groundstate density and total energy, and second, but still very important, to have an expression in which the approximation of the unknown part would be, in many cases, not as relevant as before. In the KS scheme the total energy functional is written as:

$$E[n] = T_0[n] + E_H[n] + \int d\mathbf{r} n(\mathbf{r}) v_{ext}(\mathbf{r}) + E_{xc}[n] \tag{3}$$

$T_0$ is the kinetic energy of a non interacting system with density $n$, $E_H$ is the Hartree contribution to the total energy, and $E_{xc}$ is the remaining part of the total energy which contains exchange-correlation contributions plus the difference between the kinetic energy of the non interacting system $T_0$ and the true kinetic energy $T$. Given this form for the total energy functional it can be shown that the system of interacting electrons will have the same density of a system of non interacting electrons subject to an effective potential given by $v_{KS} = v_{ext} + v_H + v_{xc}$ where $v_{xc} = \frac{\delta E_{xc}[n]}{\delta n(\mathbf{r})}$. Now for the auxiliary system of non interacting electrons a single particle Schrödinger equation can be written:

$$[-\frac{1}{2}\nabla^2 + v_{ext} + v_H + v_{xc}]\phi_i(\mathbf{r}) = \varepsilon_i \phi_i(\mathbf{r}) \tag{4}$$

so that the ground state density of the interacting system is expressed in terms of single KS orbitals:

$$n(r) = \sum_i f_i |\phi_i(\mathbf{r})|^2 \tag{5}$$

where $f_i$ is the occupation number of the state $i$. It is now possible, given an approximation for $v_{xc}[n]$, to solve the Kohn-Sham equations (4),(5) selfconsistently and hence to calculate the density of the interacting (real) system. Once the density is known, it is possible to calculate the energy of the ground state of the interacting system and hence, by proper minimisation, to find the ground state geometry of the (real) system. The simplest and most common approximation to $E_{xc}$ is the Local Density Approximation LDA given by[10]:

$$E_{xc}^{LDA} = \int d\mathbf{r} n(\mathbf{r}) \epsilon^{\text{heg}}(n(r)) \tag{6}$$

where $\epsilon^{\text{heg}}(n)$ is the exchange-correlation energy per electron of a homogeneous electron gas of density $n$. A possible way to go beyond LDA is to introduce an energy functional that depends also on the derivative of the density, the so called generalised gradient approximation (GGA) functionals[11,12]. Otherwise, as shown in ref.[13], partitioning the total energy functional in a different way with respect to the Kohn Sham decomposition given by eq. (3), one can still retain a single particle equation scheme, but exactly treat important portions of the exchange and correlation energies. In the calculation of total energies, these *generalised Kohn-Sham* schemes present the same accuracy as the original Kohn and Sham one. However, as we will discuss in more detail later, if one is interested in interpreting the KS eigenvalues as addition and removal energies, these latter schemes seem to yield results closer to experimental values.

As an example of ground state calculation we show here the results for the (111)2×1 surface of diamond. From a theoretical point of view, this surface is very interesting because the electronic structure depends strongly on the details of the geometry of its reconstruction. Experimentally, right after cleavage, the (111) surface of diamond exhibits a 1×1 periodicity. After annealing at more than 1000 K, however, hydrogen desorbs and the surface undergoes a 2×1 reconstruction. Both experiments and theory up to now agree that the reconstruction geometry is the Pandey chain model[14], but the details of the relaxation, i.e. dimerisation or buckling of the surface chains, are still not fully experimentally clarified. The exact determination of these details is important because, at the DFT level, the existence and the magnitude of the gap between surface states depend on them[15] in such a way that it is still not clear whether the experimentally observed semiconducting character of this surface is due to some geometrical deviations from the ideal Pandey chain model or, instead, to quasiparticle effects. Here we will show the results concerning the ground state geometry of this surface. Then we will see how deviations from this geometry can influence the band structure. We will employ both Kohn-Sham and generalised Kohn-Sham schemes to see if also in this case the latter yields a better agreement with experimental results. Unfortunately both schemes fail to reproduce the experimentally found semiconducting character of the surface.

The surface was modelled using a repeated slab made of 12 atomic layers of carbon atoms plus $\sim 5.4\mathring{A}$ of vacuum. The initial ionic positions were changed from the ideal position breaking all the possible symmetries of the ideal lattice except for the inversion of the z axis; all the atomic layers were allowed to move except for the central 2 layers. In the optimisation of the geometry we started from several configurations, involving Pandey chains with buckling and dimerisation. In the calculation we used different kind of functionals LDA and GGA. The final relaxed geometry, common to all the starting points, is shown in Fig.1.

We find that the buckling and the dimerisation of the chains vanish, thus

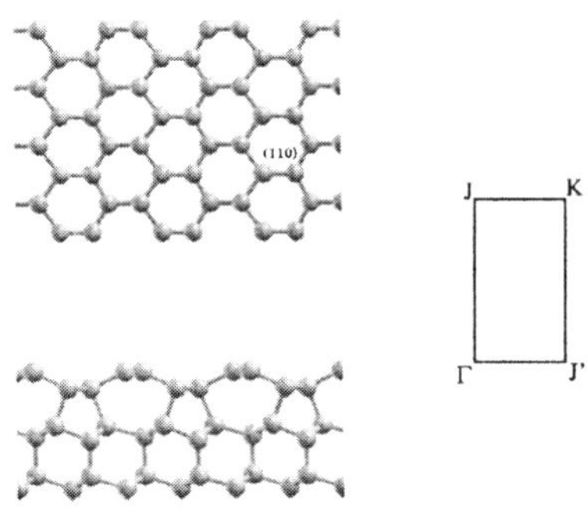


Figure 1. Pandey chain model for the (111) 2 × 1 surface of diamond. (Top view in the top panel, side view in the bottom panel). The reconstruction involves a significant change in the lattice structure. The irreducible Brillouin zone and its high symmetry points are also shown.

confirming previous DFT-LDA results[16,17].
The resulting electronic band structure is semimetallic in contrast with experimental data that predict a gap of at least 0.5 eV[18]. In particular, as we can see in Fig.2, where the DFT-GGA band structure is presented, the surface states, labelled as $N$ and $N+1$, are crossing the Fermi level along the JK line. Besides this, the agreement with the experimental dispersion of the occupied state is remarkably good.

However, it is known that the interpretation of DFT KS eigenvalues as addition and removal energies is not theoretically established[20]. The qualitative agreement with the experimental spectra is often very good but the

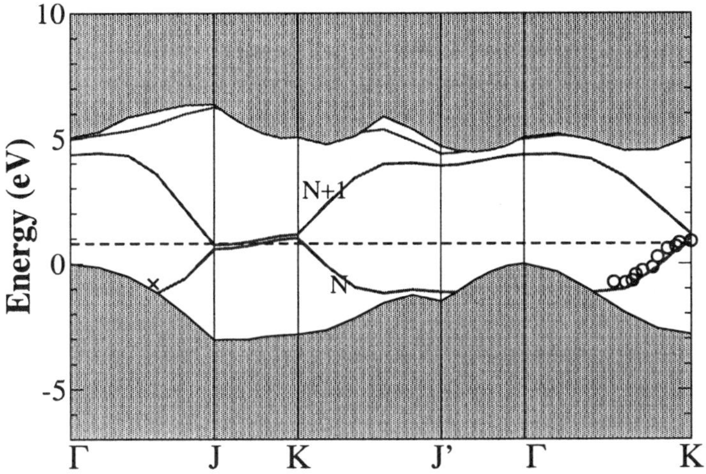


Figure 2. Electronic band structure of the 2×1 surface. In agreement with previous DFT calculation the surface is semimetallic. Between J and K the upward dispersion of the surface bands, due to the interaction between different chains, makes both the surface bands cross the Fermi level (dashed line). All the energies are referred to the top of the valence band. Circles: experiments from [18]; crosses: experiment from. [19]

band gaps are systematically underestimated. This could be the origin of the qualitative disagreement with experiment found for the C(111) surface. In fact the DFT KS gap $E_g^{DFT}$ of the $N$ particle system, is related to the *true* quasiparticle gap through the discontinuity $\Delta$ of the exchange and correlation potential when an electron is added to the system.

$$E_g = E_g^{KS} + \Delta = \epsilon_{N+1}^{(N)} - \epsilon_N^{(N)} + v_{xc}^+ - v_{xc}^-. \tag{7}$$

Where $\epsilon_{N'}^{(N)}$ represents the $N'$-th single particle Kohn-Sham level in a $N$-particle calculation. Now, if $\Delta$ is very small, a functional good enough would provide the correct gap; viceversa, if $\Delta$ is a big quantity, no matter how good the functional is, the Kohn Sham gap does not provide a good approximation for the quasiparticle gap. The discontinuity of the exchange and correlation potential was found[20] to be a consistent (80%) part of the error, the quasiparticle gap problem should then be addressed within different theories or schemes.

In the generalised Kohn-Sham schemes using hybrid functionals such as screened exchange (sX) the discontinuity of the exchange correlation potential is already partly incorporated in the single particle eigenstate of the system[13]. This fact yields reasonably good description of the electronic band structures of semiconductors, yielding gaps larger than the Kohn-Sham ones, but often smaller than the experimental gaps. As we will see in the next section, the problem of addition and removal energies can be directly addressed within many body perturbation theory (MBPT), through the introduction of quasiparticle effects, within the GW scheme[24,22,23].

In the case of the (111) surface of diamond we have tried to see if the generalised schemes would recover at least a qualitative agreement with experiments. We relaxed the surface geometry using the sX and Hartree Fock hybrid functionals. In all cases no buckling nor dimerisation was present in the final relaxed geometry, and the surface stayed semimetallic, as we can see from Fig.3

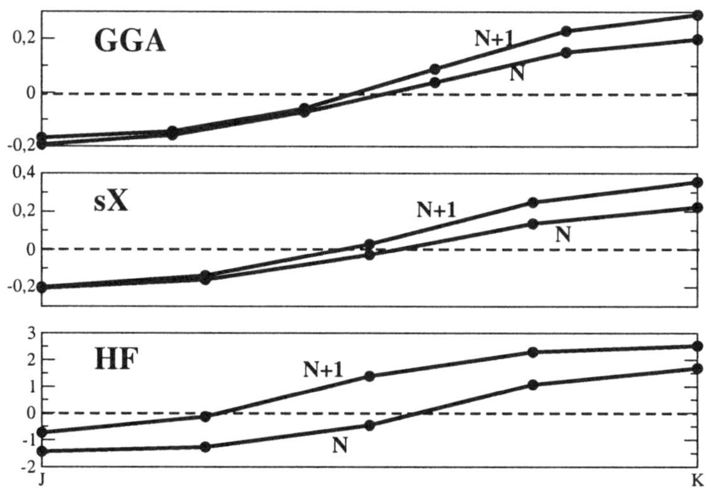


Figure 3. Surface states energies along the JK line, within the GGA, sX, and Hartree Fock schemes, for the equilibrium geometry (no buckling, no dimerisation). In all cases the surface is semimetallic. All energies are in eV and referred to the Fermi level.

## 3. Quasiparticle effects in the electronic band structure

Density Functional Theory, presented in the previous section, is a very powerfool tool to study ground state properties of electronic systems. Within this theory the Kohn Sham eigenvalues are introduced as Lagrange multipliers, so that their interpretation as excitation energies of the real system has no theoretical foundation. In fact their interpretation as excitation energies leads to the well known band gap problem of DFT: the band gaps predicted by the theory are typically 30-50% (sometimes even more) smaller than the experimental ones.

The theoretical framework that is most appropriate to treat excited state properties of many particle systems is the Green's function theory. The single particle Green's function is defined as

$$G(1,2) = -i\langle N|\mathcal{T}\{\widehat{\psi}(1)\widehat{\psi}\dagger(2)\}|N\rangle, \tag{8}$$

where $|N\rangle$ is the ground state of a $N$ particles system, $\mathcal{T}$ is the time ordering operator, $\widehat{\psi}(i)$ $(i = 1,2)$ is a field operator in the Heisenberg picture, the index $i$ represents the position $\mathbf{r}_i$, spin $\sigma_i$ and time $t_i$ coordinate, while $x_i$ includes both the position and the spin coordinates of the $i$-th particle. If we Fourier transform with respect to time we obtain the Lehmann representation of the Green's function given by:

$$G(x_1, x_2, \omega) = \sum_s \frac{f_s(x_1)f_s^*(x_2)}{\omega - \epsilon_s - i\eta\text{sgn}(\mu - \epsilon_s)}, \tag{9}$$

here $\mu$ is the Fermi energy of the system, $s$ runs over the $(N+1)$ or $(N-1)$ particles excited states and

$$\begin{cases} f_s(x_i) = \langle N|\widehat{\psi}(x_i)|N+1,s\rangle \text{ and } \epsilon_s = E(N+1,s) - E(N) \text{ when } \epsilon_s > \mu \\ f_s(x_i) = \langle N-1,s|\widehat{\psi}(x_i)|N\rangle \text{ and } \epsilon_s = E(N) - E(N-1,s) \text{ when } \epsilon_s < \mu \end{cases} \tag{10}$$

In this representation it can be seen that the poles of the Green's function are exactly the electron addition and removal energies. However, while shading light on the physical information contained in the single particle Green's function, the Lehman representation is not very useful in practice. Instead for low energies excitations and analytically continuing the Green's function in the plane of complex frequencies, a single-particle like framework can be regained introducing the concept of quasiparticles (QP). Quasi-particles can be thought as the original particles surrounded by an electron-hole polarisation cloud that screens their mutual interaction. QP are not eigenstates of the real Hamiltonian and have finite lifetimes.

The dynamic of the QPs is accounted for the self-energy operator, $\Sigma$. It is a non-local, non-hermitian, energy-dependent operator which includes the Fock exchange and all the remaining correlation effects so that QPs obey a Schrödinger-like equation similar to equations (4) of DFT:

$$H_0\psi_n(x,\omega) + \int dx'\Sigma(x,x',\omega)\psi_n(x',\omega) = E_n(\omega)\psi_n(x,\omega), \tag{11}$$

where $H_0(x_1) = -\frac{1}{2}\nabla^2 + v_{ext} + v_H$.

It can be shown that the QP equation 11 is equivalent to the Dyson equation:

$$G(1,2) = G_0(1,2) + \int d(34)\; G_0(1,3)\Sigma(3,4)G(4,2), \tag{12}$$

The Dyson equation connects the interacting Green's function $G$ to the Green's function of the system subject only to $H_0$
It can also be shown that $G$ and $\Sigma$ are fully determined by a close set of equations derived by Hedin[24] in 1965:

$$\Sigma(1,2) = i \int d(34)\; G(1,3)\Gamma(3,2,4)W(4,1^+); \tag{13}$$

$$\Gamma(1,2,3) = \delta(1,2)\delta(1,3) + \int d(4567)\; \frac{\delta\Sigma(1,2)}{\delta G(4,5)}\; G(4,6)G(7,5)\Gamma(6,7,3); \tag{14}$$

$$P(1,2) = -i \int d(34)\; G(1,3)G(4,1^+)\Gamma(3,4,2); \tag{15}$$

$$W(1,2) = V(1,2) + \int d(34)\; W(1,3)P(3,4)V(4,2); \tag{16}$$

where $1^+$ stands for $(\mathbf{r}_1, \sigma_1, t_1 + \delta)$, $\delta$ is an infinitesimal positive number, $P(1,2)$ the time ordered polarisation operator, $W(1,2)$ the dynamical screened Coulomb interaction and $\Gamma(1,2,3)$ is the vertex function.

Hedin's equations together with the Dyson's one for the Green's function form a set of equations that must be solved self-consistently. This leads in principle to the exact solution of the many-body problem, but it is practically impossible for realistic systems so that these equation are typically solved using an iterative scheme. Concerning the determination of $\Sigma$, the simplest approximation consists in starting with a non-interacting system by putting $\Sigma = 0$; in this case the Green's function is simply $G_0$, the vertex corrections are neglected with $\Gamma(1,2,3) = \delta(1,2)\delta(2,3)$ and the irreducible polarisability is given by non-interacting electron hole pairs $P(1,2) = -iG_0(1,2)G_0(2,1^+)$. With this first iteration step the self-energy becomes

$$\Sigma(1,2) = iG_0(1,3)W_0(3,1); \tag{17}$$

that is the so-called GW approximation, first introduced by Hedin[24] in 1965.
In practice, thanks to the similarity of equations (11) and (4) the problem is treated in a perturbative way with respect to the difference between the

self-energy and the Kohn and Sham exchange-correlation potential. At the first order, the quasiparticle energies are

$$E_n = \varepsilon_n + \langle \phi_n^{KS} | \Sigma(E_n) - v_{xc}^{KS} | \phi_n^{KS} \rangle. \tag{18}$$

The usual way to proceed is to expand the self-energy to first order around $\varepsilon_n$:

$$\langle \Sigma(E_n) \rangle = \langle \Sigma(\varepsilon_n) \rangle + (E_n - \varepsilon_n) \left\langle \frac{\partial \Sigma(\omega)}{\partial \omega} \right\rangle_{\omega=\varepsilon_n} + \dots \tag{19}$$

and calculate the GW corrections to the Kohn and Sham energies:

$$E_n - \varepsilon_n = \frac{\langle \Sigma(\varepsilon_n) \rangle - \langle v_{xc}^{KS} \rangle}{1 - \left\langle \frac{\partial \Sigma(\omega)}{\partial \omega} \right\rangle_{\omega=\varepsilon_n}}. \tag{20}$$

This is the standard GW approximation that typically yields quite accurate results for one-particle excitations removing the underestimation of the band gap, peculiar of DFT.

### 3.1. *Example: liquid water*

As an example of GW calculation in complex systems, we show the results obtained within this theory for a particular three dimensional system: liquid water[7]. As already mentioned, to avoid the use of very large supercell we use several MD snapshots of water as input geometries and averaged the results. We used 20 configurations of 17 water molecules in a cubic box with 15 a.u. side. We firstly used DFT to obtain Kohn-Sham eigenvalues and eigenvectors and then corrected the energy levels within the GW approximation. In the calculations we used GGA pseudopotentials and 8 k-points to sample the Brillouin zone. DFT-KS results for the electronic properties of water (obtained averaging over the 8 k points for each of the 20 configurations) are shown in Fig.4. The configuration-averaged HOMO-LUMO gap turned out to be 5.09 eV, in good agreement with previous DFT calculations [25] but strongly underestimating the experimental gap (8.7 ±0.5 eV [26]), as expected in DFT.

In order to correct the KS electronic gap we calculated the GW corrections $\Delta\varepsilon_n^{QP}$ to the KS energies. This should be done for *all* the 20 MD configuration, followed by an average. In particular the calculation of the screened Coulomb interaction for 20 configurations constitutes a true bottleneck. Instead, we observed that the GW corrections $\Delta\varepsilon_n^{QP}$ were quite stable with respect to the configuration. In other words, the difference between DFT and GW, $\Delta GW = \varepsilon_n^{QP} - \varepsilon_n^{DFT}$, was practically constant going

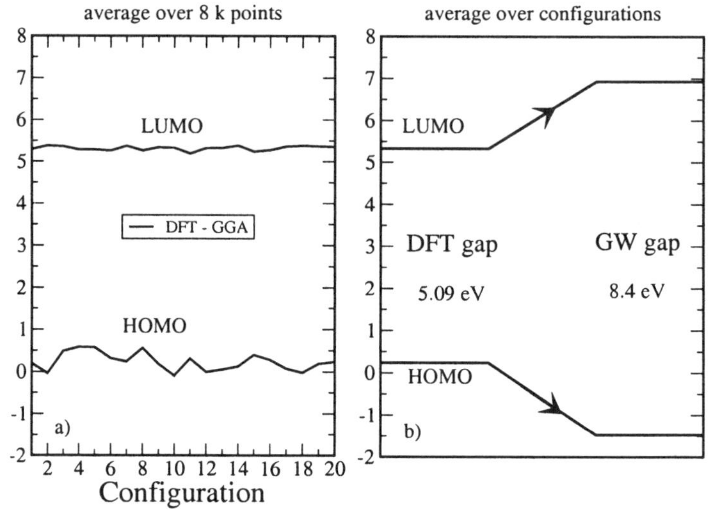


Figure 4. a) DFT HOMO and LUMO energies, averaged over the 8 k-points, for the 20 MD configurations. b) Schematic DFT and GW HOMO-LUMO gaps, averaged over the 20 MD configurations.

from one snapshot to another. This is explicitly shown in table 1 for three different configurations. We used, hence, the same GW corrections for all the DFT configurations.

Table 1. GW corrections to the HOMO and LUMO energy levels and to the GGA electronic gap, for three different water configurations (E19, E8, E2). Energies in eV.

| | DFTgap | $\Delta GW$ HOMO | $\Delta GW$ LUMO | $\Delta GW$ gap |
|---|---|---|---|---|
| E19 | 5.09 | -1.67 | 1.61 | 3.28 |
| E8 | 4.71 | -1.64 | 1.60 | 3.24 |
| E2 | 5.29 | -1.70 | 1.60 | 3.30 |

With this GW corrections the average electronic HOMO-LUMO gap is increased to 8.4 eV well within the experimental range [26], as shown in Fig.4.

## 4. Many body effects in the optical properties

The application of the approach described in the previous section yields, in many cases, to energy levels in good agreement with photoemission and inverse photoemission experiments in semiconductors, insulators and metals [27]. In this kind of experiment the system is left in a charged excited state where the electron is removed or added to system. On the other hand, in order to describe correctly the spectroscopic processes, where electron–hole pairs are created, the inclusion of solely the self-energy corrections to the DFT eigenvalues in the RPA formula for the polarisation P (which is the key quantity to obtain the measured spectra), is still not enough. In fact it is not possible to stop at the independent particle level in the description of the polarisation, but we have to introduce the electron-hole interaction (in other words a non delta-like Γ) through a second iteration of the Hedin equation [28]. This procedure allows to arrive to a Dyson equation for a generalised four point P (see ref . [28], for details of the derivation), which describes the electron-hole dynamics and reads as:

$$P(1234) = P_{IQP}(1234) + \int P_{IQP}(1256)K(5678)P(7834)d(5678) \tag{21}$$

and it is known as the Bethe–Salpeter equation (BSE). $P_{IQP}$ describes the propagation of independent quasi-electron and quasi-hole couples while all the interactions are contained in the Kernel K. The kernel K is made of two parts, deriving from the functional derivative of the Hartree potential and of the selfenergy with respect to the single-particle Green's function. So, using the GW approximation for Σ and neglecting both $\delta W/\delta G$ and the dynamical effects in $W$, the kernel becomes [28]:

$$K(1234) = \delta(12)\delta(34)v(13) - \delta(13)\delta(24)W(12). \tag{22}$$

If we write the kernel in reciprocal space and eliminate the long range divergent contribution to $v$ we can treat both the excitonic and the local fields effects at the same footing. In fact it can be shown [28] that the modified first term is taking care of the local fields effect, while the second represents the attractive screened electron-hole coulomb interaction. As we will see in the following examples the first term is particularly important in low dimensional systems, while the second one is, generally, the most important on bulk materials.

In practice, to solve the BSE, the problem is recast into an effective two-body hamiltonian form. To do this a basis of couples of LDA Bloch wavefunctions $\phi_{nk}^{LDA}$ is employed. In this basis we have that the independent particle polarisation is:

$$P^0_{(n_1,n_2),(n_3,n_4)} = \frac{(f_{n_2} - f_{n_1})\delta_{n_1,n_3}\delta_{n_2,n_4}}{\epsilon_{n_2} - \epsilon_{n_1} - \omega - i\eta}. \tag{23}$$

It can be shown that ( see ref. [29] for all the mathematical details):

$$P = [1 + iP^0K]^{-1}P^0 = [H_{exc} - I\omega]^{-1}_{(n_1,n_2),(n_3,n_4)}(f_{n_4} - f_{n_3}) \tag{24}$$

where an effective excitonic hamiltonian, defined as the following, has been introduced:

$$H_{exc}^{(n_1,n_2),(n_3,n_4)} = (E_{n_2} - E_{n_1})\delta_{n_1,n_3}\delta_{n_2,n_4} - i(f_{n_2} - f_{n_1})\times$$

$$\int d\mathbf{r}_1\, d\mathbf{r}'_1\, d\mathbf{r}_2\, d\mathbf{r}'_2\; \phi_{n_1}(\mathbf{r}_1)\, \phi^*_{n_2}(\mathbf{r}'_1)\, K(\mathbf{r}_1,\mathbf{r}'_1,\mathbf{r}_2,\mathbf{r}'_2)\, \phi^*_{n_3}(\mathbf{r}_2)\, \phi_{n_4}(\mathbf{r}'_2)$$

Hence, using the spectral representation [29] for the inverse of a matrix, the interacting polarisation can be obtained solving an effective eigenvalue problem:

$$\sum_{(n_3,n_4)} H_{exc}^{(n_1,n_2),(n_3,n_4)} A_\lambda^{(n_3,n_4)} = E_\lambda A_\lambda^{(n_1,n_2)}. \tag{25}$$

For the calculation of absorption spectra, we can limit ourselves to transitions with positive frequency[28], i.e. $(n_1, n_2)$ and $(n_3, n_4)$ are pairs of one valence and one conduction band, respectively. Moreover, we build up the spectra of optical properties by considering only negligible momentum transfer, hence the same **k** for the valence and the conduction state. In this way the macroscopic dielectric function, written in terms of the excitonic binding energies and eigenfunctions is:

$$\epsilon_M(\omega) = 1 + \lim_{\mathbf{q}\to 0} v(\mathbf{q}) \sum_\lambda \frac{\left|\sum_{v,c;\mathbf{k}} \langle v, \mathbf{k}-\mathbf{q}|e^{-i\mathbf{q}\mathbf{r}}|c,\mathbf{k}\rangle A_\lambda^{(v,c;\mathbf{k})}\right|^2}{(E_\lambda - \omega)}. \tag{26}$$

The calculation of $\epsilon_M$ is generally very demanding because the excitonic hamiltonian matrix to be diagonalised can be very large. In fact the relevant parameters which determine its size are the number of **k** points in the BZ, the number of the valence bands $N_v$, the number of conduction bands $N_c$ which build the basis set of pairs of states.
BSE calculations, performed in insulators and semiconductors [28,32,30,31], show how the inclusion of the electron-hole Coulomb interaction allows a quantitative comparison with experiments, not only below the electronic gaps, where generally bound excitons are formed, but also above the continuum edge. One of the first examples appeared in the literature is given by bulk silicon [30] where excitonic effects have been shown to enhance the E1 peak by almost 100%. Furthermore, the e–h interaction generally induces a redshift of the spectral peaks, which partially cancels the blueshift arising from the self–energy corrections. In the last ten years we assisted to a rapid growth of applications of this computational scheme to many materials from bulk semiconductors to real surfaces and nanostructures.

Here we concentrate on two applications of some silicon nanowires and on silicon nanoclusters.

### 4.1. *Examples: Silicon nanowires (1D)*

In Fig.5 we report the imaginary part of the dielectric function of a Si [100] wire with a linear cross section of 4Å for light polarised along the growth axis and perpendicular to it. These spectra are calculated for an interwire distance (minimum distance between wire surfaces), equal to 10Å, which makes the interaction among wires negligible and for this reason we will refer to them as *isolated wires.* The first row of each panel shows the optical spectrum obtained at the RPA level, confirming essentially the results obtained in ref. [33]. The second row shows how the dielectric response is modified by taking into account the inhomogeneity of the system (RPA+LF). In particular, including the so called local-field effects, we observe a small change of the optical spectrum for light polarised along the wire axis, while we find an important intensity reduction for the perpendicular light polarisation (right panel). The reason that RPA, without LF, fails for $\perp$ polarisation is due to the depolarisation effect [34] which is created by the polarisation charges induced in the system. The depolarisation is accounted only if LFs are included, and is responsible of the suppression of the low energy absorption peaks in $\perp$ direction, rendering the wire almost transparent below 8 eV. A similar anisotropic behaviour, found in other first-principles calculations on nanotubes [34,35] and nanowires [36,37], has been observed for the optical absorption of carbon nanotubes [38], for the photoluminescence spectra of porous silicon [39] and for the optical gain in silicon elongated nanodots [40].
The third row of each panel reports both the absorption spectra when self-energy corrections (GW) and self-energy, local-field and excitonic effects (GW+LF+EXC) are taken into account. For perpendicular light polarisation (right panel), the GW+LF+EXC spectrum is rather similar to the RPA+LF curve, whereas for parallel light polarisation, a big transfer of the oscillator strength to the low energy peaks is observed, together with a reduction of the Sommerfield factor above the electronic gap (as predicted in simplified models [41,42]). Large exciton binding energies, due to the larger overlap of electron and hole wavefunctions inside the wire, come out.

How does the inter-wire distance affect the optical absorption? This point is of particular importance since experimentally, very often, the individual wires are close packed and can interact with each other via long-range forces induced by excitations.
In Fig.6 are reported the theoretical optical spectra calculated at the

$GW$ and at $EXC + LF + GW$ level for different inter-wires distances: $D_{w-w} = 10\text{Å}$ top panel (isolated), $D_{w-w} = 3\text{Å}$ central panel (solid1), $D_{w-w} = 1\text{Å}$ bottom panel (solid2). Looking at right part of the three panels, we first note that the depolarisation effect decreases reducing the inter-wires distance. The packing makes the wire no more transparent for light polarised perpendicular to the axis. Moreover we observe that when the wires are very close to each other (solid2), the absorption for perpendicular light polarisation becomes more important with respect to the other polarisation. Another important effect can be seen reducing the inter-wires distance: a reduction of the opening of the DFT gaps, of the $GW$ corrections, and also a reduction of the excitonic binding energy. In particular, looking at the top and central left panels, we observe that the spectra after the inclusion of all many-body effects are quite similar. This is probably due to the lack of local fields effects and to a strong confinement of excitons within each wire, with respect to the DFT electron states. As a consequence, excitons in different wires behave quite independently for this light polarisation. Only when the inter-wire distance is further reduced to about $1\text{Å}$ (solid2), a slight red-shift of the optical $EXC + LF + GW$ spectrum starts to be visible.

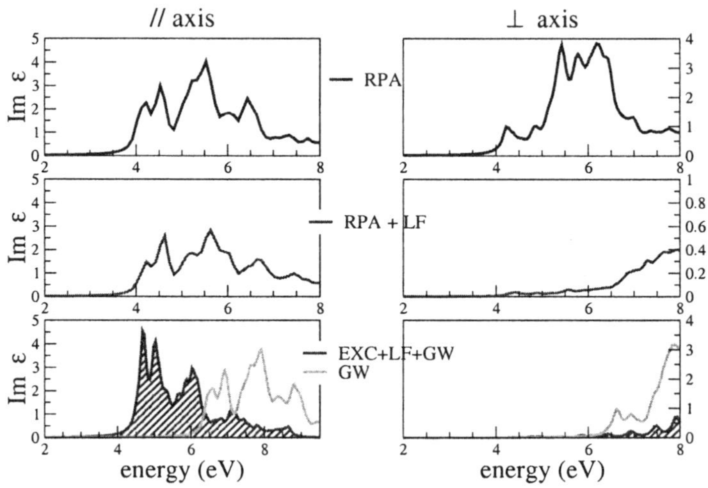


Figure 5. Imaginary part of the dielectric function of an isolated [100] SiNW (0.4 nm). Left panels: light polarised along the wire axis; right panels: light polarised perpendicularly to this axis.

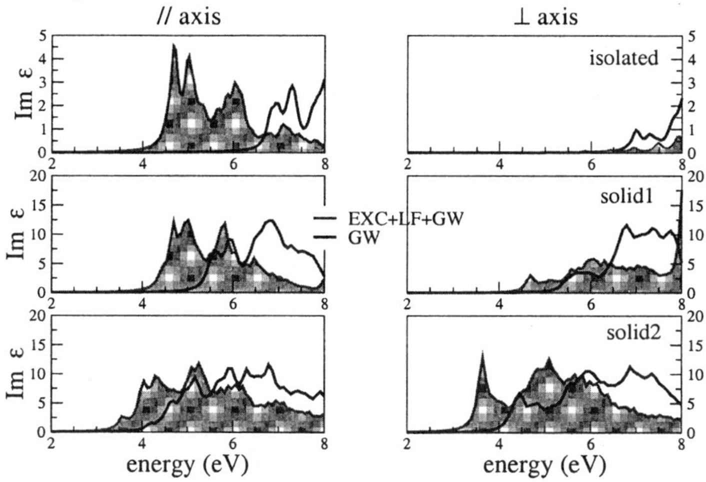


Figure 6. Imaginary part of the dielectric function of Si[100] wires: $D_{w-w} = 10\AA$ (almost isolated),a solid of interacting nanowires with $D_{w-w} = 3\AA$ (solid1), a solid of interacting nanowires with $D_{w-w} = 1\AA$ (solid2)

### 4.2. *Example: silicon nanoclusters (0D)*

Another example that shows the importance of introducing Many-Body effects in studying low dimensional systems is the emission and absorption spectra of Silicon nanocrystals (Si-nc). These have attracted increasing interest because of the possibility to engineer luminescing transitions in an otherwise indirect gap material[43]. It is generally accepted that the quantum confinement, caused by the restricted (nanometric) size, is essential for the visible light emission in Si nanostructures. Still the exact role of defects, doping, and interface structure in the photoluminescence (PL) spectra have to be clarified. In particular, the PL properties of embedded Si-nc in $SiO_2$ suggest that oxygen induces important modifications in the electronic and optical properties of silicon nanocrystals [44]. In Fig.8 we show our calculated absorption and emission spectra for a small Si nanocluster, the $Si_{10}H_{16}$, and the corresponding optical spectra in the case of $Si_{10}H_{14}O$, where two hydrogen atoms have been substituted by a oxygen atom in a bridge configuration (see Fig.7). Absorption and emission are calculated in the ground and excited states respectively, where the excited state geometry corresponds to the electronic configuration in which the highest occupied single-particle state (HOMO) contains a hole ($h$), while the lowest unoccupied single-particle state (LUMO) contains the corresponding electron ($e$).

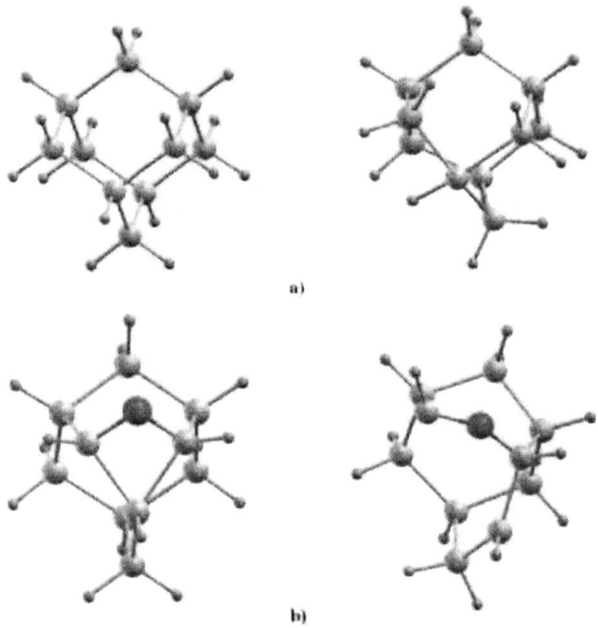


Figure 7. Geometries of a) $Si_{10}H_{16}$ and b) $Si_{10}H_{14}O$ nanoclusters. Left: ground state geometries; right: excited state geometries

Excitonic, local field and self-energy effects are included. A strong photoluminescence peak appears around 1.5 eV in the case of $Si_{10}H_{14}O$, due to a bound exciton with a large binding energy of $\sim$ 2 eV. The resulting emission spectrum well compares with the experimental spectrum of Ma et al. [44] shown in the inset. Our results suggests that the presence of a Si-O-Si bridge bond at the surface of Si-nc can explain the nature of luminescence in Si nanocrystallites embedded in $SiO_2$.

## 5. Conclusions

We have shown that parameter-free calculations of the spectroscopic properties of systems with different dimensionality and microscopic order, are now possible using Many-Body Perturbation Theory to correct the failures of the one-particle DFT approach. This scheme allows to determine both

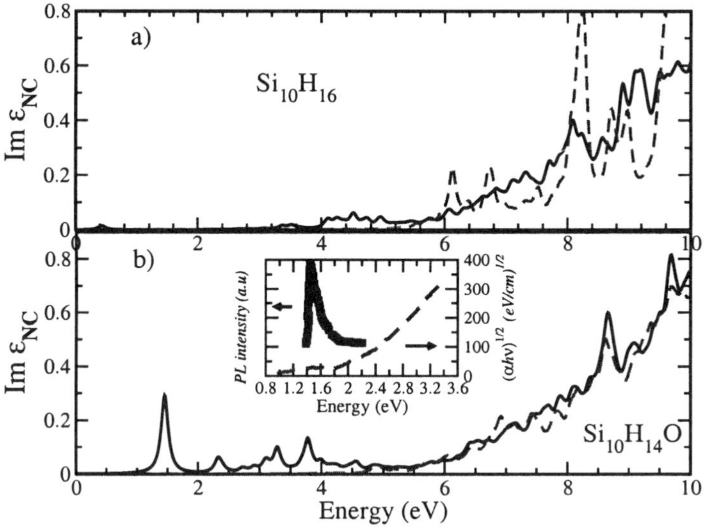


Figure 8. a) Imaginary part of the dielectric function of $Si_{10}H_{16}$ nanocluster. Solid line: emission; dashed line: absorption. b) Imaginary part of the dielectric function of $Si_{10}H_{14}O$ nanocluster. Solid line: emission; dashed line: absorption. Inset: experiment by Ma et al. [44].

ground state and excited state properties at the same level of microscopic accuracy. We have discussed self-energy effects in the GW approximation which yields quite accurate results for the electronic gaps in many materials. Moreover, we have introduced the Bethe-Salpeter equation needed to describe the electron-hole interaction and showed some examples where theoretical response are strongly influenced by many body effects.

## Acknowledgments

This work has been supported by MIUR project PRIN NANOEXC 2005, MIUR project NANOSIM, by the EU through the Nanoquanta Network of Excellence (NMP4-CT-2004-500198) and by CNISM. Computer resources from INFM "Progetto Calcolo Parallelo" at CINECA are gratefully acknowledged.

## References

1. The DFT-GGA calculations have been perfomed using the FHI98MD code [2],the PWSCF code: S. Baroni, S. de Gironcoli, A. Dal Corso, P. Giannozzi, http://www.pwscf.org/, and the ABINIT code [3] sX and HF calculations have been performed using VASP. GW and BSE calculation

have been performed using codes developed in the NANOQUANTA NoE http://www.nanoquanta.eu

2. M. Bockstedte, A. Kley, J. Neugebauer, and M. Scheffler, Comp. Phys. Comm. **107**, 187 (1997)
3. X. Gonze, J.-M. Beuken, R. Caracas, F. Detraux, M. Fuchs, G.-M. Rignanese, L. Sindic, M. Verstraete, G. Zerah, F. Jollet, M. Torrent, A. Roy, M. Mikami, Ph. Ghosez, J.-Y. Raty, D.C. Allan. Computational Materials Science 25, 478 (2002)
4. G. Onida, L. Reining, R. W. Godby, R. Del Sole, and W. Andreoni Phys. Rev. Lett. **75**, 818 (1995)
5. D. Varsano PhD Thesis, University of the Basque Country, San Sebastian, Spain, (2006)
6. C. A. Rozzi and D. Varsano and A. Marini and E. K. U. Gross and Angel Rubio, Phys. Rev. B **73**, 205119 (2006)
7. V. Garbuio et al., to appear in Phys. Rev. Lett.
8. P. Hohenberg and W. Kohn, Phys. Rev. **136**, B864 (1964).
9. R.M. Dreizler and. E.K.U. Gross *Density Functional Theory* (Springer Verlag Hedelberg, 1990) R.O. Jones and O. Gunnarsson Rev. Mod. Phys. **61**, 689 (1989)
10. W. Kohn and L. J. Sham, Phys. Rev. **140**, A1113 (1965).
11. J.P. Perdew, K. Burke, and Y. Wang, Phys. Rev. B **54**, 16533 (1986)
12. J.P. Perdew, J.A. Chevary, S.H. Vosko, K.A. Jackson, M.R. Pederson, D.J.Singh, C. Fiolhais Phys. Rev.B **46** , 6671 (1992)
13. A. Seidl, A. Görling, P. Vogl, and J.Majewski Phys. Rev. B **53**, 3764 (1996)
14. K.C. Pandey, Phys. Rev. B **25**, R4338 (1982)
15. F. Bechstedt, A.A. Stekolnikov, J. Furthmüller, P. Käckell, Phys. Rev. Lett. **87**, 16103 (2001); A.A. Stekolnikov, J. Furthmüller and F. Bechstedt, Phys. Rev. B **65**, 115318 (2002)
16. A. Scholze, W.G. Schmidt, F.Bechstedt, Phys. Rev. B **53**, 13725 (1996)
17. G. Kern, J. Hafner, J. Furthmüller, G.Kresse, Surf. Sci.**357-358**, 422 (1995)
18. R. Graupner et al., Phys. Rev. B **55**, 10841 (1997)
19. F.J. Himpsel, D.E. Eastman, P.Heimann, J.F. van der Veen, Phys. Rev. B **24**, 7270 (1981)
20. R. W. Godby, M. Schlüter, and L. J. Sham, Phys. Rev. Lett. **56**, 2415-2418 (1986)
21. L. Hedin and S. Lundqvist, in *Solid State Physics*, Ed. H. Ehrenreich, F. Seitz, D. Turnbull (Academic, New York), Vol. 23, pag.1, (1969)
22. M.S. Hybertsen, S.G. Louie, Phys. Rev. B 34, 5390 (1986)
23. R. W. Godby, M. Schlüter and L. J. Sham, Phys. Rev. B **37**, 10159 (1988)
24. L. Hedin, Phys. Rev. **139**, A796 (1965)
25. K. Laasonen et al., J. Chem. Phys. **99**, 9080 (1993)
26. see e.g. A. Bernas et al., Chem. Phys. **222**, 151 (1997) and ref. therein; http://www.ensta.fr/ muguet/papers/ECCC7/band.html and ref. therein
27. F. Aryasetiawan, O. Gunnarsson, Rep. Prog. Phys. **61**, 237 (1998)
28. G. Onida, L. Reining, and A. Rubio, Rev. Mod. Phys. **74**, 601 (2002).
29. S. Albrecht, Ph. D. thesis, Ecole Polytechnique, France (1999).
30. S. Albrecht, L. Reining, R. Del Sole, and G. Onida Phys. Rev. Lett. **80**,

4510-4513 (1998)
31. L. X. Benedict, E. L. Shirley, and R. B. Bohn Phys. Rev. Lett. **80**, 4514-4517 (1998)
32. M. Rohlfing and S. G. Louie Phys. Rev. Lett. **81**, 2312-2315 (1998)
33. A. N. Kholod, V. L. Shaposhnikov, N. Sobolev, V. E. Borisenko, F. A. D'Avitaya, S. Ossicini, Physical Review B 70 (2004) 035317.
34. A. G. Marinopoulos, L. Reining, A. Rubio, N. Vast, Phys. Rev. Lett. **91**, 046402 (2003).
35. E. Chang, G. Bussi, A. Ruini, E. Molinari, Phys. Rev. Lett. **92**, 196401 (2004).
36. M. Bruno, M. Palummo, A. Marini, R. Del Sole, V. Olevano, A. N. Kholod, and S. Ossicini, Phys. Rev. B **72**, 153310 (2005).
37. F. Bruneval, S. Botti, and L. Reining, Phys. Rev. Lett. **94**, 219701 (2005).
38. N. Wang and *et. al*, Nature **408**, 50 (2000)
39. D. Kovalev, B. Averboukh, M. Ben-Chorin, and F. Koch, Al. L. Efros and M. Rosen, Phys. Rev. Lett. **77**, 2089 (1996) (1996).
40. M. Cazzanelli and D. Kovalev and L. Dal Negro and Z. Gaburro and L. Pavesi, Phys. Rev. Lett. **93**, 207402 (2004)
41. T. Ogawa, T. Takagahara, Phys. Rev. **43** 14325 (1991) .
42. T. Ogawa, T. Takagahara, Phys. Rev. B **44** 8138 (1991).
43. S. Ossicini, L. Pavesi, F. Priolo, in *Light Emitting Silicon for Microphotonics*, Springer Tracts in Modern Physics, 194, Berlin (2003) .
44. Z. Ma, X. Liao, G. Kong, J. Chu, Applied Physics Letter 75 (1999) 1857.

# DIELECTRIC RESPONSE AND ELECTRON ENERGY LOSS SPECTRA OF AN OXIDIZED Si(100)-(2×2) SURFACE

L. CARAMELLA* and G. ONIDA

*European Theoretical Spectroscopy Facility and Physics Dept., University of Milan, via Celoria, 16, I-20133 Milan, Italy*
**E-mail: lucia.caramella@unimi.it*

C. HOGAN

*European Theoretical Spectroscopy Facility and CNR-INFM, University of Rome "Tor Vergata", via della Ricerca Scientifica, I-00133 Roma, Italy*

We compute, from first principles, the dielectric function and electron energy loss spectrum (EELS) of an oxidized Si(100)(2×2) surface. The surface local field effect is found to be important for the calculation of EELS in a reflection geometry. Theoretical problems that arise when local fields are included within the periodic supercell approach are discussed in detail.

*Keywords*: EELS; surface; response function; polarization; TDDFT; Si(100)

## 1. Introduction and motivations

Density functional theory (DFT) and time-dependent DFT (TDDFT) [1, 2] are powerful techniques for performing parameter-free calculations of dielectric response in condensed matter systems. Frequently it is found that problems arise during computation that are specific to the kind of system studied. Surfaces, for instance, require a different treatment with respect to the cases of bulk systems, isolated atoms and clusters, as their two-dimensional periodicity cannot easily be included within either a localized or extended framework. The calculation of surface excited-state properties further brings a dramatic increase in computational requirements.

In the present study, we focus on the dielectric response of the weakly oxidized Si(100)-(2×2) surface (oxygen coverage is about 1ML). Our first aim, with respect to our previous work on this surface [3], is to perform a full calculation of the dielectric matrix — i.e., to include local field (LF) effects, which were neglected therein. A second aim is to apply these results to the study of the electron energy loss spectra (EELS).

The paper is structured as follows. The theoretical background of reflectance EELS is briefly summarized in Section 2. In Section 3 we describe the system and provide some numerical details about the calculation of the dielectric function. In Section 4 we analyze the problem of interpretating the dielectric function when calculated using the supercell method for semi-infinite systems, with respect to the case of infinite and finite systems (bulk and isolated atoms or clusters respectively). Finally, we present in Section 5 some preliminary results of the EELS of oxidized Si(100)-(2×2), including local field effects.

## 2. Theory

Electron energy loss spectroscopy is one of the most important techniques for characterizing bulk solids and surfaces. In the reflection geometry, a monoenergetic electron beam strikes the sample and the backscattered electrons are analyzed with respect to energy and momentum. For low energy incident electrons (10–500eV) the penetration depth is of the order of a few Angstrom. Hence one can assume that backscattered electrons interact only with the atoms of the first layers of the crystal, and carry information about the electronic properties of the surface.

The standard dipole-scattering theory for the electron scattering probability can be found, e.g., in ref. 4. More advanced models of energy loss, that account for the anisotropy of the surface layer within a three-layer model, can be found in Refs. 5 and 6. The latter methods have been used to probe the electronic structure of several crystal surfaces at the DFT-LDA/RPA level (including GaAs(001) Ref. 12) and also including many body effects (C(100), Ref. 7).

In the present work, we assume a planar scattering, and take $yz$ as being the scattering plane ($z$ is the surface normal). The scattering probability is defined by:

$$P(\mathbf{k},\mathbf{k}') = A(\mathbf{k},\mathbf{k}')\mathrm{Im}g(\mathbf{q}_{||},\omega) \tag{1}$$

where $\mathbf{k}$ and $\mathbf{k}'$ are the incident and scattered wavevectors and $A(\mathbf{k},\mathbf{k}')$ is the kinematic factor:

$$A(\mathbf{k},\mathbf{k}') = \frac{2}{(ea_0\pi)^2}\frac{1}{\cos\theta_0}\frac{k'}{k}\frac{q_{||}}{|q_{||}^2+q_{\perp}^2|^2} \tag{2}$$

The angle $\theta_0$ represents the direction of the incident beam with respect to the normal to the surface plane, and $q_{||}, q_{\perp}$ are the parallel and perpendicular components, respectively, of the transferred momentum $\mathbf{q} = \mathbf{k} - \mathbf{k}'$.

The loss function is defined by:

$$g(\mathbf{q}_{||}, \omega) = -\mathrm{Im}\frac{1}{1 + \varepsilon_{\mathrm{eff}}(\mathbf{q}_{||}, \omega)} \tag{3}$$

Following Refs. 5 and 7 we define:

$$\varepsilon_{\mathrm{eff}}(\mathbf{q}_{||}, \omega) = \varepsilon_s \frac{\varepsilon_s + \varepsilon_b + (\varepsilon_b - \varepsilon_s)e^{-2\mathbf{q}_{||}d\varepsilon_{\mathrm{aux}}}}{\varepsilon_s + \varepsilon_b - (\varepsilon_b - \varepsilon_s)e^{-2\mathbf{q}_{||}d\varepsilon_{\mathrm{aux}}}} \tag{4}$$

where $d$ is the thickness of the surface and $\varepsilon_b$ is the bulk dielectric function. The surface dielectric functions $\varepsilon_{s,y}, \varepsilon_{s,z}$ (parallel and perpendicular respectively to the surface plane) are needed to construct $\varepsilon_s$ and $\varepsilon_{\mathrm{aux}}$, defined as:

$$\varepsilon_s = (\varepsilon_{s,y}\varepsilon_{s,z})^{1/2} \tag{5}$$

and

$$\varepsilon_{aux} = (\varepsilon_{s,y}/\varepsilon_{s,z})^{1/2} \tag{6}$$

## 3. Computational scheme and results for the dielectric function

For the ground state calculation we assume the equilibrium geometry determined in ref. 3. In Fig. 1 we show an image of the $(2 \times 2)$ cell and a lateral view of the half slab, with the reconstruction involving the formation of Si dimers, arranged in chains perpendicular to the dimer axis [13].

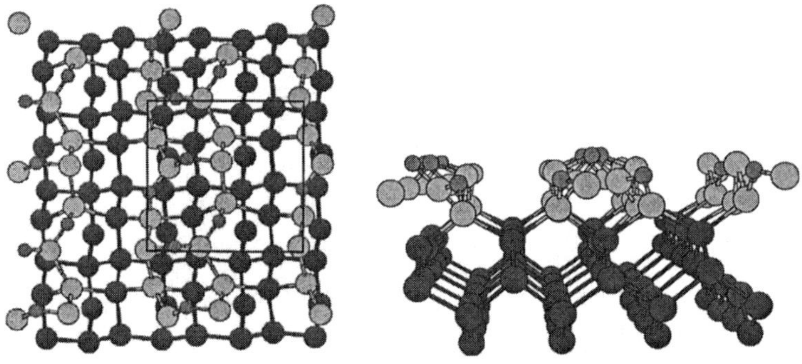

Fig. 1. Si(100)(2×2):O surface (1ML coverage). Left: top view of the surface cell. Right: lateral view of the half slab. The smaller light-grey circles represent oxygen atoms, large light-grey circles represent surface silicon atoms, while the dark grey circles are bulk silicon atoms. Dimers chains are evident in the lateral view.

We performed a self-consistent density functional theory calculation in the local density approximation (DFT-LDA [17,18]) with the ABINIT plane-wave package [16]. Kohn-Sham eigenvalues and eigenvectors were then computed on a denser reciprocal space mesh (7×7) for use as input to the DP code [10] for the *ab initio* computation of the dielectric response. We used the Hilbert transform (HT) method as described in refs. [14,15] to evaluate the dynamic independent particle response function $\chi^{(0)}(\omega)$. Then, the DP code was used to obtain the interacting response function $\chi(\omega)$ from its independent particle counterpart $\chi^{(0)}(\omega)$, and, from $\chi(\omega)$, the slab dielectric function $\varepsilon^{scell}(\omega)$. In particular, the surface dielectric functions $(\varepsilon_{s,x}(\omega), \varepsilon_{s,y}(\omega)$, and $\varepsilon_{s,z}(\omega))$ were obtained, in the $\mathbf{q} \to 0$ limit, for the three directions of polarization of incident light.

The most important modification of the imaginary part of $\varepsilon_s(\omega)$ due to the local field contributions appears for light polarized perpendicular to the surface plane, as shown in Ref. 15, because of the macroscopic discontinuity in the charge distribution. However, in the same work, the influence of the local fields on the surface optical properties was found to be small. This is not surprising since surface sensitive optical techniques such as reflectance anisotropy spectroscopy (RAS) only probe the components of the surface dielectric tensor *parallel* to the surface. The motivation of the present work should now become clear: as shown in Eq. (4), EELS spectra also require the *perpendicular* component.

## 4. The supercell approach and some related pitfalls

In this section we summarize some consequences of the conventions used to describe the dielectric functions of different systems. The supercell method is often used for the study of isolated systems: by introducing artificial periodic boundary conditions, it is possible to apply the standard tools developed for bulk crystals. The question is: how should one link *a posteriori* the calculated $\varepsilon$ for a supercell to that of a realistic surface system?

In order to answer this question it can be useful to divide the discussion into three parts according to different classes of systems: (i) bulk (infinite systems); (ii) atoms and clusters (finite systems); and (iii) surfaces (semi-infinite systems). The analysis has to start from the experimental results and the physical interpretation of the quantities calculated assuming realistic models.

In case (i), the experimental results are expressed in terms of $\mathrm{Im}\,\varepsilon_M(\omega)$ (absorption spectra) and $\mathrm{Im}\,[\varepsilon_M^{-1}(\omega)]$ (EELS). For bulk systems the microscopic dielectric function can be calculated from the following expression:

$$\varepsilon_b(\omega) = 1 - v\chi(\omega) \tag{7}$$

where $v$ is the Coulomb potential. In the case of the independent particle approximation, $\chi(\omega) = \chi^{(0)}(\omega)$.
The absorption spectrum is directly related to the response function through the *macroscopic* dielectric function:

$$\mathrm{Im}\,\varepsilon_{b,M}(\omega) = -\lim_{\mathbf{q}\to 0}\left[v_{\mathbf{G}=0}(\mathbf{q})\bar{\chi}_{\mathbf{G}=\mathbf{G}'=0}(\mathbf{q},\omega)\right] \tag{8}$$

where $\bar{\chi}_{\mathbf{GG}'}$ is the modified response function, related to the independent particle polarizability $\chi^{(0)}(\omega)$ through a Dyson equation $\bar{\chi} = \chi^{(0)} + \chi^{(0)} K \bar{\chi}$ with $K = \bar{v} + f_{xc}$ ($\bar{v}$ is the Coulomb potential without long range terms and $f_{xc}$ is the exchange-correlation term; see e.g. ref. 21). The bulk plasmon peaks, instead, coincide with the maxima of:

$$-\mathrm{Im}\left[\varepsilon_M^{-1}\right] = -\mathrm{Im}\left[v\chi\right] \tag{9}$$

In case (ii), i.e. the case of finite systems like atoms or clusters, the experimentally measured quantity is the photoabsorption cross section, and not an absorption spectrum related to Im $\varepsilon(\omega)$. The photoabsorption cross-section is given by:

$$\sigma(\omega) = \frac{4\pi\omega}{c}\mathrm{Im}\alpha(\omega) \tag{10}$$

where $\alpha(\omega)$ is the dynamic dipole polarizability [20] (e.g. the atomic polarizability or the cluster polarizability).

In the case of finite systems [19] we can still obtain Im $\alpha(\omega)$ from a calculation of $\varepsilon(\omega)$ by placing the system inside a periodic supercell of large volume $\Omega_0$ and macroscopic dielectric function $\varepsilon(\omega)$, since $\mathrm{Im}\varepsilon(\omega) = \frac{4\pi}{\Omega_0}\mathrm{Im}\alpha(\omega)$.

In the limit of an infinitely large supercell the contributions of the Coulomb potential at $\mathbf{G} = 0$ are negligible, and $\bar{v} = v$. In summary, if the local fields are included we can write:

$$\mathrm{Im}\varepsilon_M^{-1} = \lim_{\mathbf{q}\to 0}\mathrm{Im}\left[v_{\mathbf{G}=0}(\mathbf{q})\chi_{\mathbf{G}=\mathbf{G}'=0}(\mathbf{q},\omega)\right] \tag{11}$$

which means that for atoms or clusters the plasmon peaks (as observed in the EELS) coincide with the optical absorption peaks.

The case (iii) is indeed more subtle. Information about the surface comes from experiments measuring the optical reflectivity (for example in RAS or SDR) or the electron energy loss. In all cases the calculation of a dielectric function is required. However, one should obtain a dielectric function that represents the realistic response of the surface, and not the response of the specific supercell used to simulate it. Naive choices could be adopted but are not satisfactory. For instance, treating the supercell (which contains a solid slab plus some empty layers) as a bulk system would yield a dielectric function which depends on the amount of vacuum, and which approaches 1 when the vacuum increases. On the other hand, treating the supercell as an isolated system, i.e., extracting the slab polarizability from an equation similar to Equation (10), one gets a quantity which grows linearly with the thickness of the slab. This is not a problem when the slab polarizability is used to compare RAS or SDR in the independent quasi particle approach, but becomes relevant in the case of EELS calculations. Hence, we need to derive an expression for the polarizability which is independent from the geometry of the supercell. One can consider the following expression:

$$\alpha = \left( \frac{\varepsilon_M^{scell} - 1}{4\pi} \right) \left( \frac{d_s}{d_v} + 1 \right) A d_v \tag{12}$$

where $\varepsilon_M^{scell} = 1 + 4\pi\alpha/[A(d_s + d_v)]$. In this way, it is possible to derive a physically correct quantity by dividing Eq. (12) by the volume of the vacuum of the supercell and fixing the ratio $\frac{d_s}{d_v}$. In particular, if $\frac{d_s}{d_v} = 1$, then Eq. (12) is particularly simple. However, keeping $d_s = d_v$ can be computationally very expensive.

## 5. EELS spectra

The slab polarizability for the system shown in Fig. 1 and the bulk silicon dielectric function are the ingredients necessary for computing the EELS spectrum according to Eqs. 1–6. We take many-body effects into account in an approximate way, thereby correcting the well-known DFT-LDA gap underestimation, by applying a rigid shift of +0.6 eV to the conduction bands (scissor operator). Finally, we calculated the second energy derivative of the EELS spectrum in order to enhance the spectral features, as is often done in experimental works (see, e.g., ref. 8).

Our theoretical results are plotted in figure Figure 2, together with the experimental data from Ibach et al. (ref. 8) for a 1 ML oxygenated Si(100) surface. Noticeably, the inclusion of local fields yields better agreement with experiment, as the peak at 5.0 eV is well described. This feature corresponds to the $E_2$ peak of bulk silicon and is also present in EELS spectra of clean Si surfaces (see, e.g., Ref. 9). Similarly, a shoulder (a peak in the RPA results) is also seen in the theoretical spectrum around 3.6 eV that coincides with the $E_1$ structure in the experimental data.

However, the broad oxygen-related peak at 7.2 eV (it is absent for clean surfaces) is less well reproduced, as the calculation with local fields predicts a relatively small peak at about 6.8 eV. This may be due to the fact that the chosen reconstruction (1 ML, with oxygen on both Si dimers and Si-Si backbonds) is not alone sufficient to represent the structure of the real oxidized surface (see ref. 23). Further investigation is in progress to characterize better the oxygen-related peaks, both in the low-energy range (below 3 eV) and in the region of the surface and bulk plasmons (about 9 and 16 eV, respectively) [22].

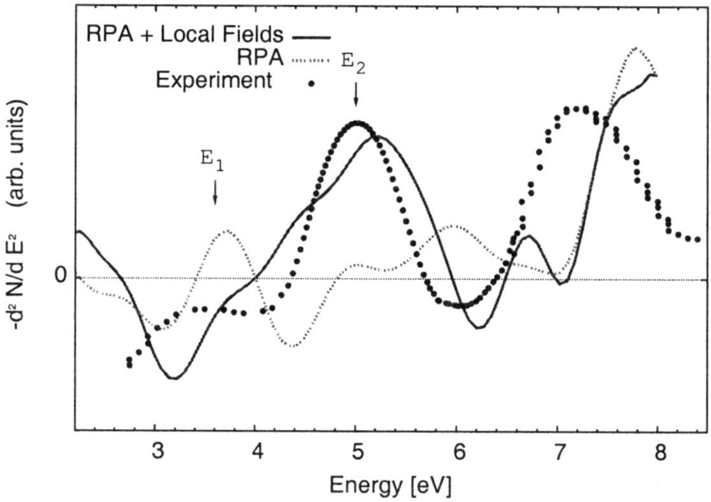


Fig. 2. Calculated second energy derivative of the electron energy loss spectrum of Si(100)(2×2):O (1ML), plotted in arbitrary units, including local field effects (full line, LFE). Dashed line: RPA results without LFE. Dots are the experimental data from ref. 8.

## 6. Conclusions

We have applied an efficient numerical method to the calculation of the dielectric response of the Si(100)(2×2):O surface, computing the EEL spectrum of a 1ML-oxidized surface up to about 8.0 eV, including the local field effects. The latter are found to improve the agreement with experimental spectra in the considered energy range. Problems associated with defining the response function in supercell calculations were also discussed in detail. Future work will include the calculation of the EELS spectrum in a wider energy range and a systematic comparison of clean and oxidized surfaces, in order to study oxygen-related peaks above 10 eV.

## Acknowledgments

This work was supported by the *Ministero Italiano dell'Università e della Ricerca* and by the EU's 6th Framework Programme through the NANOQUANTA Network of Excellence NMP4-CT-2004-500198. G.O. gratefully acknowledge CNISM for financial support. We thanks the *Institut des Nanoscience de Paris* (INSP) for the computer facilities.

## References

1. E. Runge, E. K. U. Gross, *Phys. Rev. Lett.* **52** , 997 (1984).
2. M. A. L. Marques, E. K. U. Gross, *Annu. Rev. Phys. Chem.* **55** , 427 (2004).
3. A. Incze, G. Onida, R. Del Sole, *Phys. Rev. B* **71** , 035350 (2005).
4. H. Ibach and D. L. Mills, in *Electron Energy Loss Spectroscopy and Surface Vibrations* (Academic Press, New York, 1982).
5. A. Selloni and R. Del Sole, *Surf. Science* **168** , 35 (1986).
6. R. Esquivel-Sirvent and Cecilia Noguez, *Phys. Rev. B* **58** , 7367 (1998).
7. M. Palummo, O. Pulci, A. Marini, L.Rening, R. Del Sole, *preprint* (2006).
8. H. Ibach and J. E. Rowe, *Phys. Rev. B* **9** , 1951 (1974).
9. H. H. Farrell, F. Stucki, J. Anderson, D. J. Frankel, G. J. Lapeyre, M. Levinson, *Phys. Rev. B* **30** , 721 (1984).
10. `http://theory.polytechnique.fr/codes/codes.html`
11. D. J. Chadi, *Phys. Rev. Lett.* **43** , 1 (1979).
12. A. Balzarotti, M. Fanfoni, F, Patella, F. Arciprete, E. Placidi, G. Onida, and R. Del Sole, *Surface Science* **524** , 71 (2003); E. Placidi, C. Hogan, F. Arciprete, M. Fanfoni, F. Patella, R. Del Sole, and A. Balzarotti, *Phys. Rev. B* **73**, 205345 (2006).
13. J. Fritsch and P. Pavone, *Surface Science* **344** , 159 (1995).
14. T. Miyake, F. Aryasetiawan, *Phys. Rev. B* **61** , 7172 (1999).
15. L. Caramella, G.Onida, F. Finocchi, L. Reining, F. Sottile, *submitted to Phys. Rev. B* (2006).
16. `www.abinit.org`
17. W. Kohn and L. J. Sham, *Phys. Rev. A* **140** , 1133 (1965).
18. P. Hohenberg and W. Kohn, *Phys. Rev. B* **136** , 864 (1964).
19. M. Gatti and G. Onida, *Phys. Rev. B* **72** , 045442 (2005).
20. M. J. Stott and E. Zaremba, *Phys. Rev. A* **21** , 12 (1980).
21. G. Onida, L. Reining, A. Rubio, *Rev. Mod. Phys.* **74** , 601 (2002).
22. L. Caramella, C. Hogan, G. Onida, *in preparation* (2007).
23. A. Incze, G. Onida, R. Del Sole, M. Fuchs *et al.*, *in preparation* (2007).

# DIELECTRIC FUNCTION OF THE Si(113)3×2ADI SURFACE FROM AB-INITIO METHODS

Katalin Gaál-Nagy* and Giovanni Onida

*Dipartimento di Fisica and ETSF, Università degli Studi di Milano, via Celoria 16, I-20133 Milano, Italy* * *E-mail: katalin.gaal-nagy@physik.uni-regensburg.de*

We have investigated the imaginary part of the dielectric function Im($\epsilon$) of the (113) 3×2 ADI reconstructed surface of silicon. The calculations have been performed for a periodic slab within the plane-wave pseudopotential approach to the density-functional theory. The three diagonal components of Im($\epsilon$) have been derived from the momentum matrix elements within the independent particle random phase approximation (IPRPA). In this article, the importance of the **k**-point convergence is figured out by inspecting **k**-point resolved spectra.

*Keywords*: ab initio calculations, optical properties, high-index silicon surfaces

## 1. Introduction

The Si(113) surface is one of the most stable high-index surfaces of Si [1]. It is used as a substrate for the self-assembled growth of Ge nanowires and Ge as well as SiGe islands[2–5]. Since this surface is atomically smooth, ultrathin oxide films can be grown on Si(113)[6,7], which is hence also dealt as a candidate for wafers and devices.

The clean surface shows a 3×2 periodicity at room temperature which can transform into a 3×1 one by a phase transition induced by high temperatures or contamination[8,9]. The most probable is the Si(113)3×2ADI (adatom-dimer-interstitial) reconstruction, which was proposed by Dąbrowski et al.[10] (see Fig. 1). Comparing ab-initio results for reasonable surface models, the ADI reconstruction has the lowest surface energy[11]. Besides, an experimental confirmation is given by STM measurements[4,12]. However, still other models can not be ruled out[13].

A complementary study of the Si(113)3×2 surface can be done by the investigation of its optical properties, e.g., the reflectance anisotropy spectra (RAS). At the aim of performing a theoretical ab-initio investigation of the optical properties the first step is the calculation of the imaginary part of

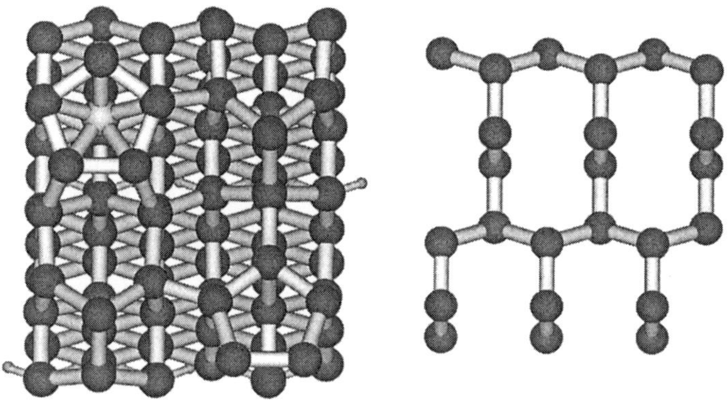

Fig. 1. In the left panel, a top view ($x$-$y$ plane) of the Si(113)3×2ADI reconstruction is shown. The right panel displays a top view of just two "double layers" of the bulk-like part of the slab. The Si atoms are dark (the interstitial Si atom light) and the hydrogen atoms light.

the dielectric function Im($\epsilon$). Results for Im($\epsilon$) are presented in this work.

## 2. Method

The calculations have been performed within the plane-wave (PW) pseudopotential approach to the density-functional theory[14,15] as implemented in ABINIT[16] and TOSCA[17]. For the exchange-correlation energy the local-density approximation has been used[18,19]. The ground state convergence required 4 **k** Monkhorst-Pack[20] points in the irreducible wedge of the Brillouin zone (IBZ) as well as a kinetic-energy cutoff of 12 Ry ($\approx$ 24000 PW). The slab used for the calculation contains 11 double layers (DL) of Si where the bottom surface is saturated with hydrogen atoms (see Figs. 1 and 4). The topmost 4 DL have been relaxed, and the remaining have been kept to bulk positions.

The imaginary part of the dielectric function Im[$\epsilon(\omega)$] as a function of the energy $\omega$ has been derived by a summation of the matrixelements of the momentum operator[21]. The sum has to be taken over the valence- and the conduction states (here: 175 and 185, respectively), and weighted **k** points, which in the present case can be taken just in one quarter of the Brillouin Zone, which is the IBZ of the system. For the layer-by-layer analysis modified matrixelements have been used according to the prescriptions of reference 22 (see also references 23 and 24). Here, the effects of local fields, self energy, and excitons have been neglected.

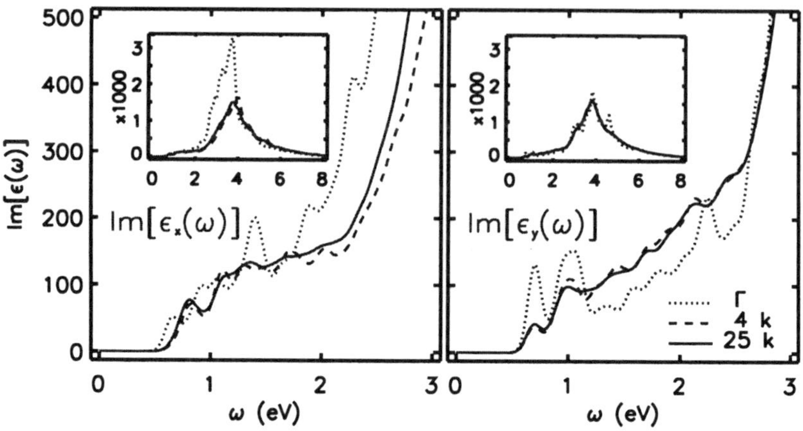


Fig. 2. Im[$\epsilon_x(\omega)$] (left) and Im[$\epsilon_y(\omega)$] (right) for various numbers of **k** points (assignment in the graph) for the low-energy range, where the surface states are found. The full energy range is shown in the insets.

## 3. k point convergence

For the calculation of Im[$\epsilon(\omega)$] (and consequently, for the calculation of all optical properties which are based on this) the convergence with respect to number of **k** points in the summation described above is essential, since the spectra are particularly sensitive to that. Working out of convergence can yield wrong difference spectra (e.g., RAS). In comparison, the convergence with respect to the number of bands is less demanding in the surface-related low-energy range. Thus, we performed a very careful check of the **k**-point convergence.

Since the slab was chosen with the surface perpendicular to $z$, Im[$\epsilon(\omega)$] should be investigated for light polarized in $x$ and $y$ direction, since these functions are the ingredients for difference spectra like the RAS. We show in Fig. 2 the Im[$\epsilon_x(\omega)$] and Im[$\epsilon_y(\omega)$] for various sets of **k** points. For both polarizations, at least 25 **k**-points are required. The convergence for Im[$\epsilon_y(\omega)$] is faster than for Im[$\epsilon_x(\omega)$]. In particular, the differences between 4 and 25 **k** points for the Im[$\epsilon_x(\omega)$] is still significant.

In order to improve the **k**-points grid in the most efficient way, we have performed a single-**k** analysis. This means, we have inspected the contributions of each **k** point to the spectra separately, which are shown in Fig. 3. It is apparent that the changes for the Im[$\epsilon_x(\omega)$] from one **k** to a neighbouring one, especially in $x$ direction, are larger than for Im[$\epsilon_y(\omega)$]. Thus, we have chosen 30 additional **k** points in between along the $x$ direction. The spectra

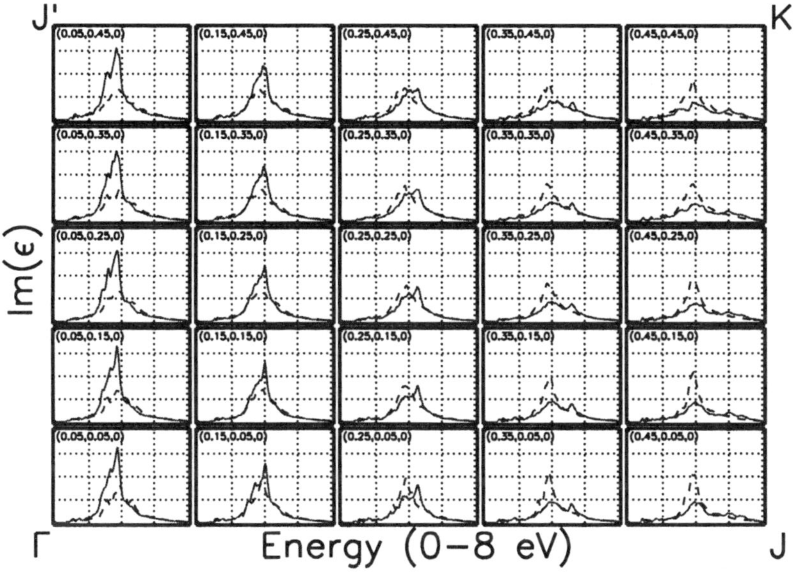


Fig. 3. Single-**k** decomposition of the spectra using 25 **k** in the surface IBZ. $\mathrm{Im}[\epsilon_x(\omega)]$ is drawn with solid and $\mathrm{Im}[\epsilon_y(\omega)]$ with dashed lines. The coordinates of the **k** points are denoted in the inset. The spectra are organized as the **k** points in the surface IBZ, where the Γ point is close to the low-left corner of the figure.

based on the resulting 55 **k** points (not shown in Fig. 2) demonstrates, that convergence already had been achieved with 25 **k** points. This strategy allows one to avoid a "blind" increase of the **k**-point grid, which can be very time-consuming.

## 4. Anisotropy at Γ

In Fig. 2 a particular feature is visible: in the contribution coming from the Γ point, a strong anisotropy at the medium-high energy range, between the $x$ and the $y$ polarization, occurs. This anisotropy at Γ appears in the energy range of bulk-bulk transitions. In fact, performing a layer-by-layer analysis the spatial localization of such anisotropic contributions is clearly found to be in the bulk (internal) DLs, as shown in Fig. 4, together with a side view of the slab cell where the DLs used for each spectra are marked. We have chosen always pairs of DLs, e.g., L07L08 corresponds to the spectra with contributions from DL 7 and DL 8, where the DLs are counted from the top to the bottom. Of course, bulk silicon is isotropic, which results also from calculations using a sufficient amount of **k** points. The bulk anisotropy at

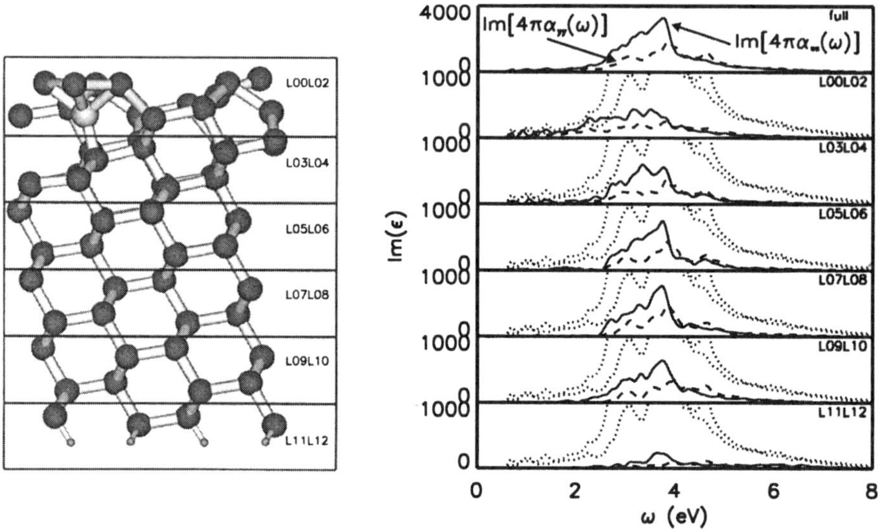


Fig. 4. Layer-by-layer decomposition of the spectra and complete spectra containing all layers (full) for the Γ point calculation (right). The spectra correspond to the layers denoted in the inset and marked at the structure (left). $\mathrm{Im}[\epsilon_x(\omega)]$ is drawn with solid and $\mathrm{Im}[\epsilon_y(\omega)]$ with dashed lines. In addition, the full-slab spectra are drawn with dotted lines. DL L00 corresponds to the vacuum at the top of the slab and DL12 to the hydrogen at the bottom.

the Γ point here is just due to the use of a low-symmetry supercell.

Fig. 4 shows that this anisotropy is present at each pair of DLs in the bulk region of the slab, where the spectra for bulk region, i.e. L09L10, L07L08, and L05L06, are nearly the same. The explanation can be found by inspecting the geometry of one pair of DLs only, which is shown on the right in Fig. 1. An highly anisotropic "chain structure" oriented along the $x$ axis can be seen. This is the reason for the strong signal appearing at Γ (the wavefunctions are in phase and the wavelength of light is much larger than the cell dimension).

## 5. Summary

In summary, we have presented ab-initio results for the imaginary part of the dielectric function for the Si(113)3×2ADI surface. At this example we have shown the possibility of accelerating the convergence tests with respect to **k** points by performing a single-**k** analysis. We have applied a layer-by-layer analysis allowing us to explain a strong anisotropic contribution coming from the Γ point. The results described here are useful for further calculations, e.g., the RAS spectra, which can be obtained from $\mathrm{Im}[\epsilon(\omega)]$.

## 6. Acknowledgment

This work was funded by the EU's 6th Framework Programme through the NANOQUANTA Network of Excellence (NMP4-CT-2004-500198). K.G.-N. also likes to acknowledge S. Hinrich and A. Stekolnikov for support.

## References

1. D. J. Eaglesham, A. E. White, L. C. Feldman, N. Moriya and D. C. Jacobson, *Phys. Rev. Lett.* **70**, p. 1643 (1993).
2. H. Omi and T. Ogino, *Phys. Rev. B* **59**, p. 7521 (1999).
3. M. P. Halsall, H. Omi and T. Ogino, *Appl. Phys. Lett.* **81**, p. 2448 (2002).
4. Z. Zhang, K. Sumitomo, H. Omi, T. Ogino and X. Zhu, *Surf. Interface Anal.* **36**, p. 114 (2004).
5. M. Hanke, T. Boeck, A.-K. Gerlitzke, F. Syrowatka and F. Heyroth, *Appl. Phys. Lett.* **86**, p. 223109 (2005).
6. H.-J. Müssig, J. Dąbrowski and S. Hinrich, *Solid-State Electron.* **45**, p. 1219 (2001).
7. H.-J. Müssig, J. Dąbrowski, K.-E. Ehwald, P. Gaworzewski, A. Huber and U. Lambert, *Microelectron. Eng.* **56**, p. 195 (2001).
8. C. C. Hwang, H. S. Kim, Y. K. Kim, K. W. Ihm, C. Y. Park, K. S. An, K. J. Kim, T.-H. Kang and B. Kim, *Phys. Rev. B* **64**, p. 045305 (2001).
9. K. Jacobi and U. Myler, *Surf. Sci.* **284**, p. 223 (1993).
10. J. Dąbrowski, H.-J. Müssig and G. Wolff, *Phys. Rev. Lett.* **73**, p. 1660 (1994).
11. A. A. Stekolnikov, J. Furthmüller and F. Bechstedt, *Phys. Rev. B* **67**, p. 195332 (2003).
12. J. Knall, J. B. Pethica, J. D. Todd and J. H. Wilson, *Phys. Rev. Lett.* **66**, p. 1733 (1991).
13. K. S. Kim, J. U. Choi, Y. J. Cho and H. J. Kang, *Surf. Interface Anal.* **35**, p. 82 (2003).
14. P. Hohenberg and W. Kohn, *Phys. Rev.* **136 B**, p. 864 (1964).
15. W. Kohn and L. J. Sham, *Phys. Rev.* **140 A**, p. 1133 (1965).
16. `http://www.abinit.org`.
17. `http://users.unimi.it/etsf`.
18. J. P. Perdew and A. Zunger, *Phys. Rev. B* **23**, p. 5048 (1981).
19. D. M. Ceperley and B. J. Alder, *Phys. Rev. Lett.* **45**, p. 566 (1980).
20. H. J. Monkhorst and J. D. Pack, *Phys. Rev. B* **13**, p. 5188 (1976).
21. R. DelSole, Reflectance spectroscopy - theory, in *Photonic Probes of Surfaces*, ed. P. Halevi (Elsevier, Amsterdam, 1995).
22. C. Hogan, R. DelSole and G. Onida, *Phys. Rev. B* **68**, p. 035405 (2003).
23. C. Castillo, B. S. Mendoza, W. G. Schmidt, P. H. Hahn and F. Bechstedt, *Phys. Rev. B* **68**, p. R041310 (2003).
24. P. Monachesi, M. Palummo, R. DelSole, A. Grechnev and O. Eriksson, *Phys. Rev. B* **68**, p. 035426 (2003).

# MODELING OF HYDROGENATED AMORPHOUS SILICON (a-Si:H) THIN FILMS PREPARED BY THE SADDLE FIELD GLOW DISCHARGE METHOD FOR PHOTOVOLTAIC APPLICATIONS

A.V. SACHENKO[1], A.I. SHKREBTII[2*], F. GASPARI[2], N. KHERANI[3] and A. KAZAKEVITCH[2]

[1] *V. Lashkarev Institute of Semiconductor Physics NAS, Kiev, Ukraine*
[2] *Faculty of Science, University of Ontario Institute of Technology, Oshawa, ON, Canada*
[3] *Electrical & Computer Engineering, University of Toronto, ON, Canada*

We present results of combined theoretical and experimental study of the thin hydrogenated amorphous silicon (a-Si:H) films based solar cells. The films for efficient and inexpensive solar cells were grown by the Saddle Field Glow Discharge Method. An analytical model to optimize photoconversion efficiency a-Si:H based solar cells with contact grid has been developed. This two-dimensional model allows an optimization of the $p^{+}$-$i$-$n$ sandwich in terms of carrier mobilities, layers thickness, doping levels and others. The geometry of the grid fingers that conduct the photocurrent to the bus bars and ITO/$SiO_2$ layers has been optimizes as well as the effect of non-zero sun incidence angles. We demonstrate the optimization method to the typical a-Si:H solar cells.

## 1. Introduction

Thin film hydrogenated amorphous silicon (a-Si:H) is widely used for photovoltaic applications.[1] Amorphous silicon based solar cells (SC) are very promising because of low production cost, possibility of covering large uneven areas, and sufficiently high efficiency. In order to get the best possible performance of the a-Si:H solar cells it is important to (i) produce a high quality amorphous films and (ii) optimize the films and solar cells in terms their parameters such as, for instance, *p*-, *i*- and *n*-layer doping levels, electron and hole mobilities $\mu_n$ and $\mu_p$ and their lifetime, resistance of *p*-, *i*- and *n*-layers, contact grid geometry and parameters of the transparent conducting and antireflecting layers.

Two of the authors (F.G. and N.K.) have developed the so-called dc saddle-field glow discharge technique (see, [2] and references therein). This efficient technique was used to deposit the hydrogenated thin film for the amorphous solar cells under consideration.

A few approaches have been proposed to optimize the performance of various types of solar cells[1]. Advantages of the analytical models are in being physically intuitive and predictive, possibility of quick and accurate estimation of photo-conversion efficiency. Two of the authors (A.V.S. and A.I.S.) have developed analytical two-dimensional models for optimization of the crystalline solar

* Corresponding author, e-mail "Anatoli.Chkrebtii@uoit.ca"

cells[3]. The one-dimensional analytical model of photoconversion efficiency for amorphous silicon based solar cells has been proposed by one of the author (A.V.S.)[4] Although analytical one-dimensional photoconversion models have been successfully developed in the past,[4,5] numerical methods are being mainly used recently to optimize the solar cell, especially those amorphous Si based.[6]

In this paper we discuss the growth and characterization of the amorphous a-Si:H films, obtained by saddle-field glow discharge method first. Next we illustrate the application of the efficient analytical model of the photoconversion in the a-Si:H solar cells that we developed. Detailed description of the theoretical approach will be presented elsewhere.[7]

## 2. Experiment: techniques and film growth

DC Saddle-Field Glow Discharge: The dc saddle-field glow discharge technique is based on the idea of oscillation of electrons in a symmetric dc electric field that effectively extends the mean free path of electrons at low pressures beyond the dimensions of the vacuum chamber.[2] The dc electric field is created in a volume between a central anode, which is semitransparent (for example, we use a wire grid), and two cathodes both of which are parallel to the anode and are symmetrically positioned on either side of the anode (see Fig. 1). As a result, the density of high energy electrons between the two cathodes is increased many fold over a conventional dc discharge, and the probability of impact ionization and other forms of activation of gas phase species is significantly increased.

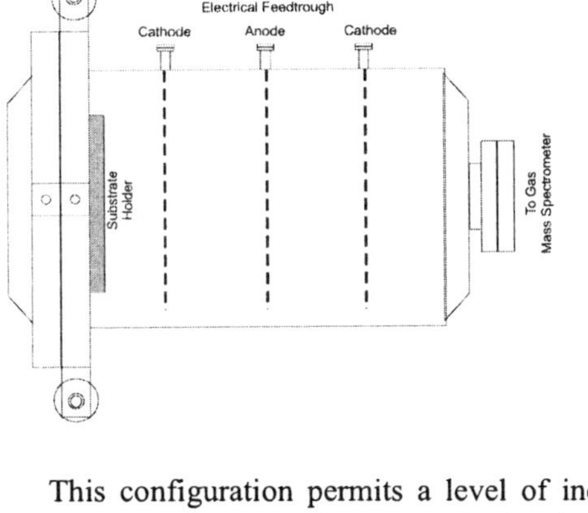


Fig. 1. DC saddle-field deposition system. The grounded substrate holder is heated and the substrate bias allows manipulation of the charge on the surface of the substrates.

This configuration permits a level of independent control over the plasma parameters on the one hand and the local substrate environment, namely substrate ion currents and ion energies, on the other hand, offered until now only by the electron cyclotron resonance technique. The saddle-field method allows deposition of the high quality amorphous films at comparatively low temperature (for instance 100°C), thus enabling the use of the low cost substrates. It scales easily in two dimensions as well and therefore allows the uniform exposure of large area substrates. Finally, the saddle-field plasma, which is robust, remote, and scaleable for large area production, is quite economic to implement.

The thin films of hydrogenated amorphous silicon studied in this paper were grown at a temperature of 250°C, 200°C, 150°C and 100°C. Their thicknesses were in the range of 200 and 400 nm. The preparation conditions and optoelectronic characterization are described in.[8] Visible and ultraviolet spectroscopies, isothermal capacitance transient spectroscopy (ICTS), Fourier transform infrared spectroscopy (FTIR), and constant photocurrent method (CPM) were used for film characterization. The films were highly resistive ($\sigma_d$ = $10^{-8}$ $\Omega^{-1}$cm$^{-1}$), and significantly photosensitive (photo to dark electrical conductivities ratio $\sigma_{ph}/\sigma_d >$ $10^4$). An optical gap $E_g$ of the films grown (called a mobility gap as well) was in the range of 1.7-2.0 eV while the gap value depends on the deposition temperature, gas pressure, substrate bias, and flow rate. Activation energy $E_a$, of the amorphous films varied from 0.6 eV to 0.9 eV.

It was previously demonstrated that such films can be successfully used for photovoltaic application[8]. However, to get the best possible performance of the a-Si:H solar cells it is important to optimize them in terms of the of amorphous film properties, *p, i* and *n* layer doping levels and conductivity, carrier mobility $\mu_h$ and $\mu_p$ and their lifetime, contact grid geometry, etc. Finally, the solar cells might not be optimally oriented with respect to the sun and additional optimization with respect to the angle of the light incidence is required as well.

## 3. Theoretical model and discussion

We consider a conventional SC of a sandwich type (see Fig. 2). Parameters of the a-Si:H film were taken from the above measurements.

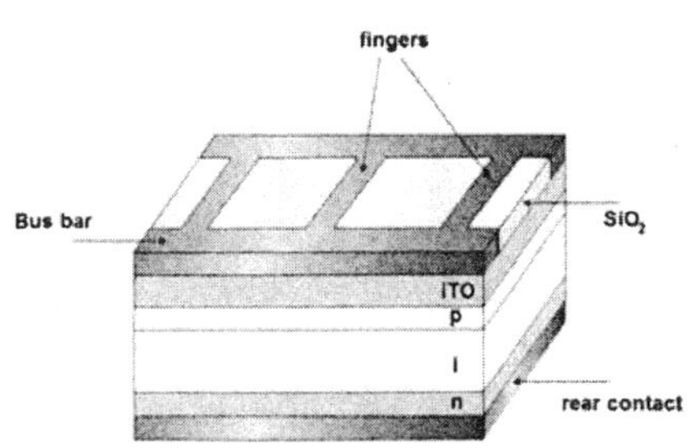


Fig. 2. Schematic view of the solar cell under consideration. It consists of the $p-i-n$ structure between frontal grids and rear contact. The frontal collecting grid electrode contains parallel metal fingers connected to each other through two conductive bus bars. The frontal grid is placed on a top of transparent conducting ITO/$SnO_2$ layer.

The optimization was performed in two steps. The first step optimizes the *p-i-n* junction parameters, especially thickness of *i*-layer. The second step optimizes the top photocurrent collecting grid and transparent conducting Indium-Tin-oxide (ITO) and Silicon oxide ($SiO_2$) layers and their thicknesses to be anti-reflecting.

For the theoretical calculation of the solar cell efficiency it is important to choose a most accurate model of photocarriers motion and collection. Hydrogenated amorphous silicon is characterized by reasonably low electron and hole mobility compared to the case of crystalline silicon. Therefore, for such amorphous material the diffusion theory of photo-conversion is more appropriate. (In contrast, the diode theory is typically applied for the crystalline Si solar cells[7].) In addition, we considered a possibility of non-zero sun incidence angles. If the solar cell is not perpendicular to the solar flux, additional optimization of solar cells is required. Such optimization was performed as well[7].

We demonstrate now the application of the formalism developed. The most important parameters of solar cells are short circuit current density $J_{sc}$, the open circuit voltage $V_{oc}$ and their efficiency $\eta$. We have found that for maximizing $J_{sc}$ in the a-Si:H solar cells the optimization of i-layer thickness is crucial. In addition, as high as possible reflection from the rear contact is required to maximize the amount of light absorbed in the relatively thin amorphous layer. Although we used $R_d$=0.9 for most of our theoretical results, it can vary significantly in practice. Diffusion length of the photoexcited carriers $L$ is another important parameter that has to be included in the optimization. The interplay between the film parameters is demonstrated in Fig. 3. This shows the calculated density of the short circuit current $J_{sc}$ with respect to *i*-layer thickness $d$ for different rear contact reflection coefficients $R_d$.

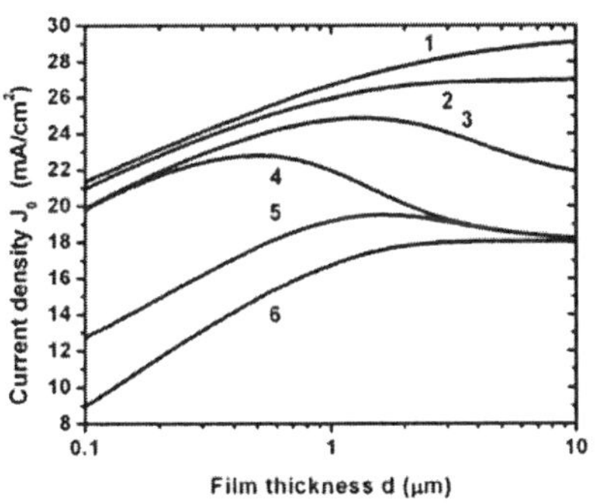


Fig. 3. Photocurrent density dependence on the thickness of *i*-layer $d$. Curve 1 corresponds to the maximum possible photocurrent density, when all the light is adsorbed in the film. Curves 2 - 6 correspond to the photocurrent $J_{sc}$ for reflection coefficients $R_d$ from the rear contact equal to 0.8, 0.6, 0.4, 0.2 and 0 respectively

This figure demonstrates gradual and monotonic increase of maximum photocurrent $J_{\lim}(d)$ with *i*-layer thickness *d* (curve 1). When the rear contact is not reflective ($R_d$ = 0), the density of the short circuit current $J_{SC}(d)$ increases with *d* and saturates after 1 μm (curve 6). Curves from 2 to 5, located between $J_{\lim}(d)$ at $R_d$=1 and $J_{SC}(d)$ at $R_d$ = 0, correspond to $R_d$ values of 0.8, 0.6, 0.4 and 0.2. The current density reaches its maximum when $d_m \approx L$ and increases when the bulk recombination velocity $S_d$ decreases. Finally, $J_{SC}(d)$ approaches $J_{\lim}(d)$ when the diffusion length $L$ is much larger than $d$ and surface recombination $S_0$ as well as $S_d$ can be neglected.

The next step was the optimization of the front contact grid parameters. This was based on the concept of effective collection length, developed in[3]. The photoconversion efficiency $\eta$ has been calculated with respect to the metaliza-

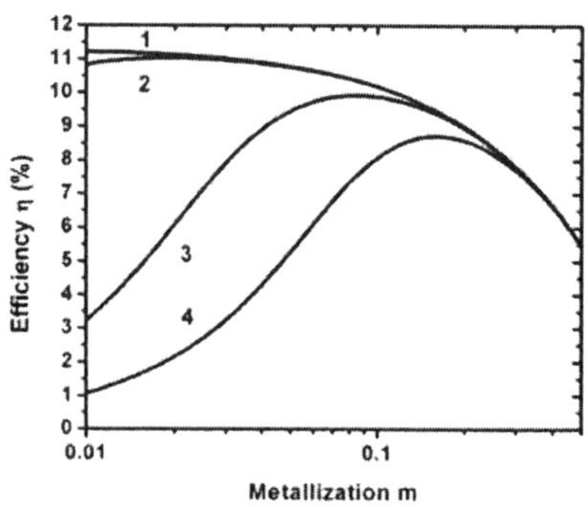


Fig. 4. Dependence of the solar cell efficiency $\eta$ on metallization coefficient *m*. The following parameters were used: $D$=2.5·10$^{-3}$ cm$^2$/s, $R_d$ = 0.9; E=10$^5$ V/cm; $R_s$= 0.082; $J_s$ = 10$^{-12}$ A/cm$^2$; $A$=1.5; $T$=300 K; $\mu_e$= 0.1 cm$^2$/V·s; $\mu_p$=102 cm$^2$/V·s; N=10$^{15}$ cm$^{-2}$. Values of $J_s$ and A correspond to parameters of high-efficient SC. Curves 1,2,3 and 4 corresponds to the metal finger thickness of 0.003, 0.01, 0.1 and 0.3 cm respectively

tion coefficient that is the shadowing of the photoactive area by the metal grid. For instance, for narrow metal contacts ($m \ll 1$) their resistance becomes a limiting factor and efficiency $\eta$ decreases. For high *m* value ($m \sim 1$) the contact grid prevents the light absorption in *p-i-n* structure, which decreases $\eta$ as well. As a results, $\eta(m)$ should show a maximum. This is demonstrated in Fig. 4, considering highly efficient a-Si:H solar cells with non-ideality coefficient $A$=1.5.

Finally, light induced degradation of performances of the a-Si:H solar cells (Staebler-Wronski effect[1]) increases the density of states inside the mobility gap. As we have found, this effect reduced the efficiency of the a-Si:H solar cells to 7.5%. Detailed discussion on the Staebler-Wronski effect contribution to the a-Si:H efficiency reduction is given in.[7]

## 4. Conclusions

Low temperature grown a-Si-H amorphous films have been optimized for photovoltaics application. We have found that the thickness of the base (*i* – layer) is a key parameter is to be optimized first for the highest efficiency. The formalism optimizes the geometry of the top metal contact grid as well. Maximum conversion efficiency that can be achieved (before Staebler-Wronski effect) is 11%, while Staebler-Wronski effect decreases efficiency to 7.5%.

## Acknowledgements

The research was supported by the Centre for Materials and Manufacturing/ Ontario Centres of Excellence (OCE/CMM) "Sonus/PV Photovoltaic Highway Traffic Noise Barrier" project.

## References

1. "Amorphous Silicon Based Solar Cells," X. Deng and E. A. Schiff, in *Handbook of Photovoltaic Science and Engineering*, A. Luque and S. Hegedus, editors (John Wiley & Sons, Chichester, 2003), p. 505 – 565
2. T. Kosteski, N.P. Kherani, F. Gaspari, S. Zukotynski and W.T. Shmayda, J. Vac. Sci. Technol. A, **16**, 893 (1998)
3. A.V. Sachenko, A. I. Shkrebtii. Ukrainski Fiz. Zhurn. (Ukrainian J. of Phys.) **29**, 1855 (1984) (in Russian). Recent implementation of the formalism is available online in English: A.V. Sachenko, A.P. Gorban. Semiconductor Physics, Quantum Electronics & Optoelectronics, **2** (2), 42 (1999) (http://www.journal-spqeo.org.ua/n2_99/42_299.htm)
4. A.V. Sachenko, Soviet Phys. Semicond., **19**, 903 (1985)
5. T. Tiedje, Appl. Phys. Lett. **40**, 627 (1982)
6. J. Lianga, E. A. Schiff, S. Guha, B. Yan, and J. Yang, Appl. Phys. Lett. **88**, 063512 (2006)
7. A.V. Sachenko, I. O. Sokolovskyi, A. Kazakevitch, A.I. Shkrebtii, submitted
8. D. Yeghikyan, N.P. Kherani, T. Kosteski, B. Bahardoust, I. Milostnaya, T. Allen†, S. Zukotynski. Proceedings of the 21st Photovoltaic Solar Energy Conference, 7-11 June 2004, Paris, France

# HIGH SPATIAL RESOLUTION RAMAN SCATTERING FOR NANO-STRUCTURES

E. SPEISER[1], B. BUICK[1], S. DELGOBBO[1], D. CALESTANI[2] AND W. RICHTER[1]

[1] *INFM, Dipartimento di Fisica Università di Roma "Tor Vergata", Via della Ricerca Scientifica 1, I-00133 Roma, Italy*

[2] *IMEM-CNR, Parco Area delle Scienze 37/A, 43010 Parma, Italy*

Nano structures like nanotubes, nanorods and quantum dots are in the center of interest in many areas of solid state physics. However, most analytical techniques do not have a sufficient spatial resolution in the nanometer-range and integrate over many structures unless their density can be reduced substantially. This holds especially true for optical techniques. In this article we describe how optical measurements with nanometer spatial resolution can be performed. Our focus is on Raman scattering but the optical equipment, exploiting the optical near field in an A-SNOM (Aperture less Scanning Near field Optical Microscope) configuration serves as well for other optical techniques like photoluminescence, transmission or reflection.

## 1. Introduction

One of the most current and also most promising field of research in solid state physics is the one of nanostructured materials. In particular, there is a great interest in nanostructured semiconductors, thanks to the latest developments in preparation methods like MOVPE (Metal Organic Vapor Phase Epitaxy), MBE (molecular beam epitaxy), lithography and colloidal chemistry. These techniques allow to prepare nano-sized semiconductors with excellent crystalline structure and most often also with epitaxially determined orientations with respect to the template. The interest in nanostructures is triggered by the fact that their physical properties, due to confinement, are different from those of the corresponding bulk material. Consequently, the size becomes a new design parameter. The latter opens a wide potential outlook to technological applications such as light emitting devices, photovoltaic cells and single electron transistors.

For 2-dimensional nanostructures (thin layers, quantum wells), being already prepared more than two decades by the standard epitaxial techniques (MBE, MOVPE), most of the standard surface science techniques could be applied. As a consequence there is excellent knowledge about their physical properties and accurate comparison with theory. However, to obtain measurements from 1-dimensional (quantum wires, nanorods, nanotubes) and 0-dimensional nanostructures (quantum dots), the additional requirement occurs that the probe must provide lateral resolution in the nanometer range. This requirement excludes or limits many of the standard surface science techniques for their analysis and is especially true for the standard optical tools with diffraction limited spatial resolution just in the sub-micrometer range. If possible, the parameters of the growth process of the nanostructures in question are modified in order to reduce the spatial density of the nanostructures in a way that only one structure is contained in the probing area. However, this creates an artificial or at least different growth process and is not always possible.

On the other hand there is a great need for optical and spectroscopic analysis of nanostructures, as has already been demonstrated for bulk semiconductors, in order to determine the electronic and vibrational properties. Because of confinement effects they will be different from those of the bulk and thus open the possibility to test the fundamental theoretical aspects as well as envision new technological applications.

Of course sufficient spatial resolution is clearly provided by scanning probe microscopes (SPM) and especially the STM (scanning tunnelling microscope) [1] is an extremely valuable tool to study nanostructures. Thus already some time ago the SPM technique was adapted to a SNOM (Scanning Near field Optical Microscope) where the sub-wavelength resolution obtainable using optical glass fiber tips with a small aperture ($< 100nm$) was exploited [2]. Spatial resolution in the nm-range was obtained.

However, for optical spectroscopy this solution has the essential drawback. The intensity that is transmitted trough the aperture depends strongly on its size. In practice to obtain sufficient signal levels a reasonable large aperture size should be chosen. On the other hand for nanometer range resolution a smaller aperture size is needed. Consequently a compromise is necessary to obtain sufficient resolution at reasonable signal to noise ratio [3, 4]. Thus, seeking for improved spatial resolution one has a serious signal to noise ratio problem. For Raman scattering which is our main interest here, as an higher order optical process, this means that only few successful experiments have been performed [5, 6].

A few years ago, however, an aperture less version of the SNOM named A-SNOM has been presented which does not have these intensity problems [7]. It worked with a sharp silicon tip on a AFM (atomic force microscope [8]) cantilever. In addition, when the tip is made from materials supporting surface plasmons in the spectral range of interest, this can lead to a strong electric field enhancement which can be very much localized, depending on the apex radius (sharpness) of the tip. The A-SNOM technique using this enhancement effects has been applied also quite recently to Raman scattering with either a modified AFM with a metallized cantilever [9, 10] or with a sharp silver tip [11] for enhanced signals from molecules and has been termed in that case Tip Enhanced Raman Scattering (TERS) .

## 2. Raman Scattering

There exist many reviews on Raman scattering in solids dealing with the basic features of this process which combines two photons (incident and scattered) and an elementary excitation of the solid [12–14]. Resonant phenomena, especially important for nano structures with their small scattering volume, have been especially treated in Ref. [15, 16]. Specific considerations for low dimensional structures are discussed in Ref. [17]. Therefore we discuss here only the main features of Raman scattering.

In the Raman scattering process a certain amount of energy is gained or lost by an incident photon with energy $\hbar\omega_i$ (*incident*) in order to create or annihilate elementary excitations of the solid, usually phonons, resulting in a scattered photon of a different energy $\hbar\omega_s$ (*scattered*). The amount of energy transferred corresponds to the eigenenergy $\hbar\omega_j$ of the elementary excitation involved labeled by "$j$":

$$\hbar\omega_s = \hbar\omega_i \pm \hbar\omega_j. \tag{1}$$

Here the "minus" sign stands for a phonon excitation (Stokes process) while the "plus" sign implies a phonon annihilation (anti-Stokes process).

The momentum $\vec{k}_j$ transferred to the vibrational excitation is related to the momentum of the incident $\vec{k}_i$ and scattered light $\vec{k}_s$ according to:

$$\hbar\vec{k}_s = \hbar\vec{k}_i \pm \hbar\vec{k}_j. \tag{2}$$

It is quite small in periodic structures when compared to the maximum possible quasi-momentum at the edge of the Brillouin Zone. Thus one usually assumes $\hbar\vec{k}_j \approx 0$. In low dimensional solids, when the periodicity is reduced in one or more directions, of course the momentum conservation is relaxed in the reduced dimensions because the wave vector is not a good quantum number any more. Instead one should discuss the phonons in those directions as confined modes. From the theoretical side however not much has been published in that respect except for the 2-dimensional superlattices [12, 18].

Of course the scattering intensity is experimentally important. This can be expressed as dipole radiation using a generalized dielectric susceptibility $\tilde{\chi}(\omega_i, \omega_s)$ which is also often called Raman tensor [19]:

$$I_s = I_i \frac{\omega_s^4 V}{(4\pi\epsilon\epsilon_0)^2 c_0^4} |\vec{e}_s \epsilon_0 \tilde{\chi}(\omega_i, \omega_s) \vec{e}_i|^2 \tag{3}$$

where $I_i$, $I_s$ and $\vec{e}_i$, $\vec{e}_s$ denote the intensity and polarization, respectively of incident and scattered light.

In a microscopic quantum mechanical approach the generalized susceptibility may be described using time-dependent perturbation theory [13]. The dominant term amounts to [20]:

$$\chi_{\alpha,\beta}(\omega_i, \omega_s) = \frac{e^2}{m_0^2 \cdot \omega_s^2 \cdot V} \sum_{e,e'} \frac{\langle 0|p_\alpha|e'\rangle\langle e'|H_{E-L}|e\rangle\langle e|p_\beta|0\rangle}{(E_{e'} - \hbar\omega_s)(E_e - \hbar\omega_i)} \tag{4}$$

where $m_0$ is the electron mass, $V$ the scattering volume, $p_\alpha$, $p_\beta$ the Cartesian components of the momentum operators, $E_e$, $E_{e'}$ the energies of the excited electron-hole pair states and $H_{E-L}$ the electron-phonon interaction Hamiltonian.

Equ. (4) includes the transition from the ground state $|0\rangle$ to an excited electronic state $|e\rangle$ (photon absorption), scattering of the generated electron-hole pair into another state $|e'\rangle$ via electron-lattice interaction, and finally the transition back to the electronic ground state $|0\rangle$ under photon emission.

The important feature of these two equations is first of all that they describe scattering selections rules. In Equ. (3)certain tensor components of the generalized dielectric susceptibility $\tilde{\chi}(\omega_i, \omega_s)$ are selected by the polarisation unit vectors $\vec{e}_i$, $\vec{e}_s$ of the light fields. From Equ. (4)they are less visible but contained in the dipole matrix elements. Secondly in Equ. (4) the brackets in the nominator shows that the intensity can be strongly increased when the photon energies match the transition energies of electron-hole pair states. Thus for a given material with given electronic states the choice of laser line energy for the scattering experiment is one of the most important parameters. In the following we describe how Raman scattering measurements on nanostructures can be performed. First, in the next section 3 with diffraction limited spatial resolution (Frauenhofer) as so called ***Micro-Raman*** scattering and afterwards in section 4 the near field spatial resolution Raman scattering so called ***Nano-Raman*** scattering.

## 3. Micro - Raman Scattering on Nanostructures

In Micro Raman scattering a microscope is utilized to focus the laser light on the sample and to collect the scattered light. In order to push the resolution to the diffraction limit the microscope is usually of the confocal type (pinholes at intermediate focal points and a laser supplying a $TEM_{00}$ mode) thus resembling as much as possible a point light source [21]. In the theoretical description of the optical microscope, the image of point sources is described by Airy functions (*Point-Spread-Function*) and the functional shape of the main maximum is used to obtain the well known result for the resolution in the Frauenhofer (far field) limit [22, 23]. The Airy disc radius, $\Delta r$ , which corresponds to the Abbe definition of the resolution limit, is given by the distance between the maximum and the first minimum of the Airy function [21]:

$$\Delta r = 0.6098 * \frac{\lambda}{NA}, \tag{5}$$

where $\Delta r$ corresponds to the distance between two just resolvable light sources, $\lambda$ is the light wave length and NA= $\sin\alpha$ is the light collecting numerical aperture at the collection angle $\alpha$.

The side maxima of the Airy function, however, are thereby neglected. Consequently the Confocal Microscope utilizes pinholes to suppress them and increases thereby also the contrast.

### 3.1. *Raman Experimental Set-up*

We discuss the experimental configurations with the example of the set-up used at Roma "Tor Vergata" (Fig.1). Its different configurations allow for measurements on different spatial resolution scales: macro-, micro- and nano-Raman spectroscopy. Common to all resolution scales is the laser illumination and the spectroscopic unit. By directing the laser light into different, parallel existing optical arrangements, the different spatial resolutions can be chosen.

The illumination source common for all set-ups is an argon-ion laser with a very low beam divergence (0.5 mrad) emitting mainly a $TEM_{00}$ mode. The mode purity is further enhanced by expanding the beam and by applying spatial filtering.

The spectroscopic tool common to all three resolution configurations is a triple monochromator that can be operated in single line scanning mode (added dispersion) and in a multichannel mode as a spectrograph. The scanning mode provides a higher contrast and resolution when working with a low dark count photomultiplier. In the multichannel mode the fist

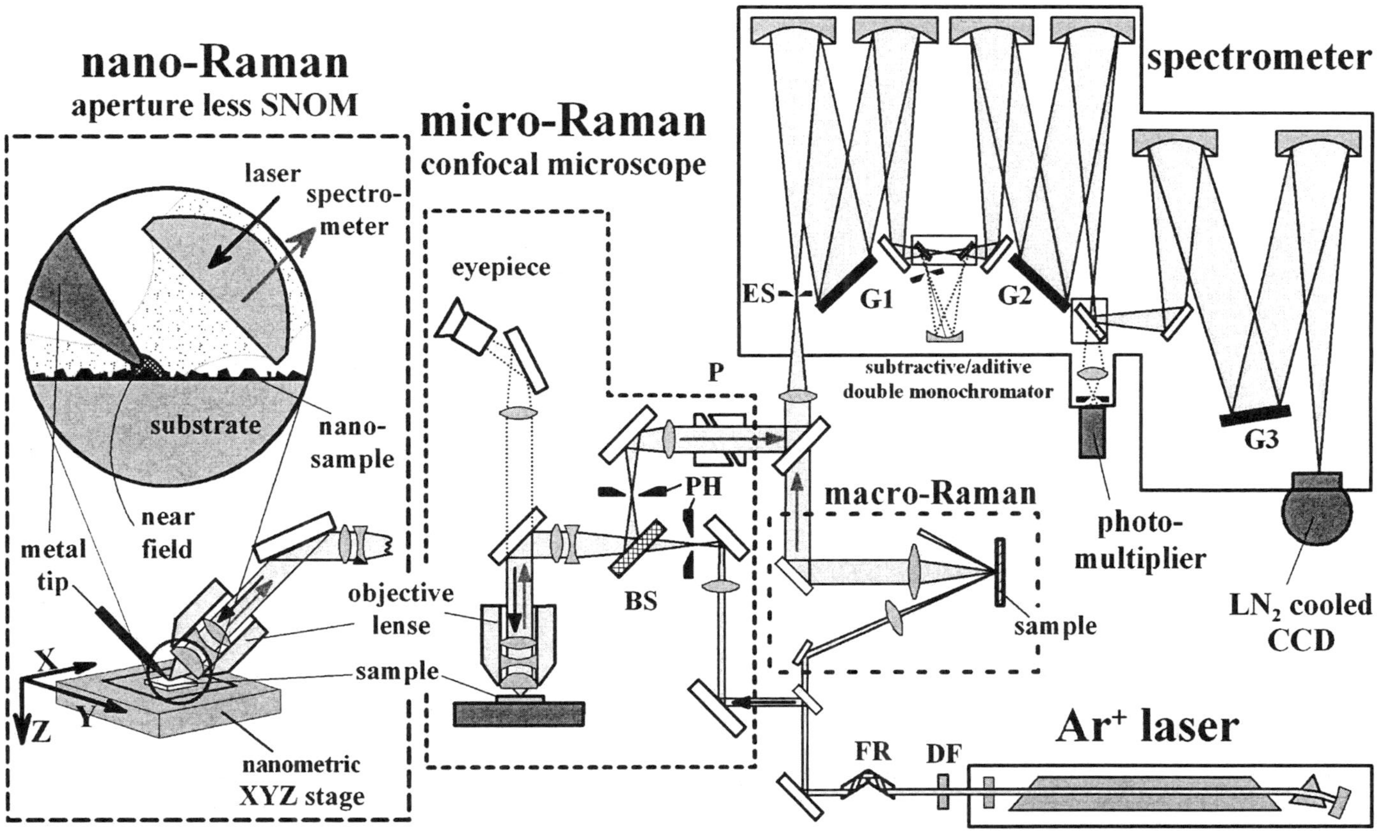


Figure 1. Raman setup with macro-, micro and nano-optics. **BS**-beam splitter; **FR**-Fresnel rombs; **PH**-pin holes; **P**-polarizer prism; **DF**-dielectric laser line filter; **ES**-monochromator entrance slit; **G1,G2G3**-holographic gratings.

two monochromators operate in a subtractive mode, selecting the desired spectral region, to be analyzed by the spectrograph equipped with a low a noise liquid nitrogen cooled CCD (charge coupled detector).

The macro-Raman configuration, as the standard Raman set-up is mainly used when the experiment needs long working distances. It allows only the use of small numerical apertures (NA = 0.1-0.2) as it is often the case with in-situ measurements in ultrahigh Vacuum chambers or gas cells. The resolution is thus low (equ. 5) around $10 - 100\mu m$. The micro-Raman set-up is based on a confocal microscope design containing an objective with a very large numerical aperture (NA = 0.95) and a pinhole in the intermediate focal plane. The spatial resolution of this microscope is below 1 $\mu$m. In nano-Raman where the optical near field is exploited resolutions below 100 nm are achieved. Independent of the desired resolution, it turned out to be extremely important in all measurements on nanostructures, to have nano-positioning capabilities (nanometric stages) for the sample, in order to be able to increase the optical signal.

### 3.2. *Micro-Raman measurements on Nano structures*

In Micro-Raman the diffraction limited spatial resolution of several 100 nm is best exploited by confocal optics. The illuminated region includes in general several nanostructures being randomly distributed and oriented due to the growth procedures (Fig. 2). In general the nanostructures in the laser focus will have different properties due to size, shape and orientation (Fig.2b). Thus deriving well defined properties of single nanostructures from the measurement averaging over several different structures is difficult. Even if a single nanostructure can be isolated the entire nanostructure will contribute to the signal. Hence its internal structure e.g. hetero-junctions or inhomogeneities (defects, stoichiometry) still remain indistinguishable. Furthermore, only a small part of the laser light excites the nano-sample. Thus severe intensity and signal-to-noise ratio problems in single nanostructure measurements might be encounted. Consequently micro-Raman has been applied up to now mainly to nano materials with a large Raman cross section.

Despite of these drawbacks the information on the phonon frequencies can still be extracted and quite a large number of data have been published. This concerns first of all data on carbon nanotubes [24–28]. We present in this section some examples performed on the 1-d nanostructures shown in Fig.2. As indicated by the white circles, which indicate micro-Raman spatial resolution, the spectra represent averages over several nanostructures.

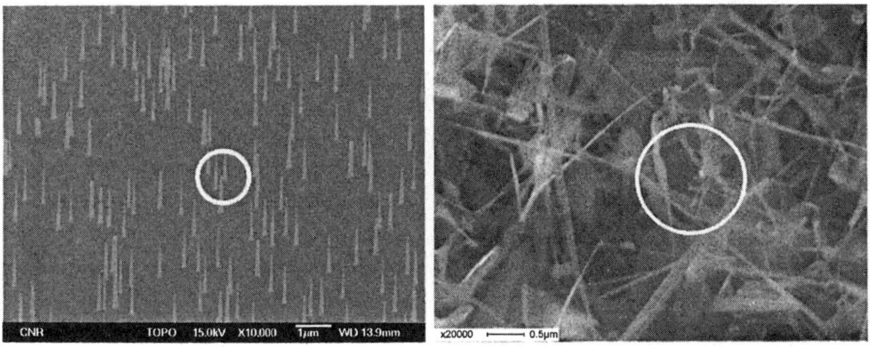


(a) (b)

Figure 2. Nanosamples from which spectra are shown in this section. The white circles represent the spatial resolution of the micro-setup: a)GaAs nanorods grown on GaAs(111) substrate by the VLS growth process [29]. The diameter of the nanorods ( 60nm) corresponds essentially to the size of the gold nano-droplets deposited on the substrate for nucleation purposes. (b)ZnO nanowires grown by thermal evaporation. The size and the shape of the nano-structures vary in a wide range dependent on the position in the sample.

### 3.2.1. *GaAs nanowires*

GaAs nanorods (Fig.2a) grown by the Vapor Liquid Solid (VLS) method [30] in (111) direction on GaAs(111) substrates in an oriented fashion were obtained from the University of Lecce [29]. They have diameters of approximately 60 nm and lengths in the micrometer range. The measurements were performed on the samples as they were grown. Single nanorods were selected in the low nanostructure density area of the sample and positioned into the micro-focus by moving the sample on a nanometric stage. In order to distinguish the Raman signal of the nanorods from the Raman scattering from that of the substrate (both nanorods and the substrate are crystalline GaAs) the focal plane of the confocal microscope was placed 0.5-1 $\mu$m above the substrate surface.

The spectra of the GaAs nanorods show a strong dependence on the applied laser power due to the heating of the rods by the laser beam [31]. Already at comparably low laser power density (several mW per $\mu m^2$) the small volume of the nanostructures can be heated easily up to several hundred K above room temperature. As a result shifts of the phonon frequencies to lower values and a line shape broadening are observed similar as observed for Si nanowires [17]. The anharmonicity of the lattice potential in GaAs [32, 33] is the cause for these modification with laser power. Thus nano-rods turn out to be a very convenient laboratory for anharmonic

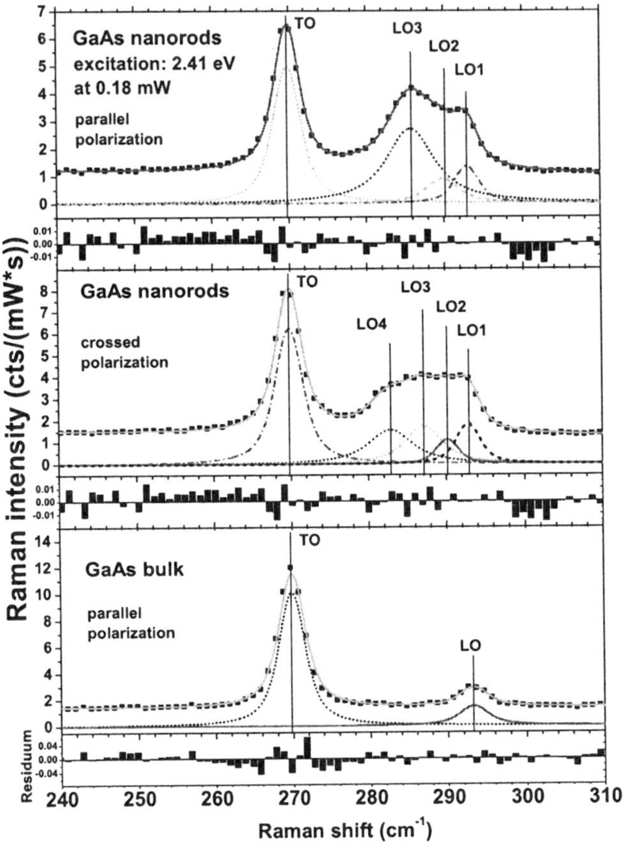


Figure 3. Raman spectra from a single free standing GaAs nanorod (two upper panels) and GaAs bulk (lower panel). The three spectra are fitted by Voigt line shapes in order to verify quantitative frequency shifts. No significant frequency shifts of the LO and TO phonons from the nanorods in respect to the bulk could be verified. In comparison to bulk spectra new lines appear in the nanorod spectra in the gap between TO and LO phonons. The new phonon lines are sensitive to the polarization of the excitation laser light as can be seen from the comparison of the upper and middle panel spectra.

lattice dynamics. In case low dimensional effects are investigated, the influence of the heating should be avoided and only low excitation laser power can be used. In the measurements shown in Fig.3 a laser power of 180 $\mu W$ from the 2.41 eV laser line was used after a series of measurements with lower and higher powers .

From the low power Raman spectra of the GaAs nanorods (Fig.3) it was found that the TO phonon frequency corresponds to that of the bulk. This is expected since the diameter of the nanorods (60 nm) is not sufficiently small to result in a quantum confinement of the phonon modes. But for the LO phonons a remarkable change in the appearance of the phonon line is clearly visible in the spectra. Several new modes appear in the TO-LO gap and dominate the LO region. A simple explanation for their appearance would be that the macroscopic LO phonon electrical field acts at longer distances (phonon wave length) and already "feels" the boundaries even in these rather large structures. In contrast, the TO phonons, arising from the short range crystal potential forces, would need smaller dimensions to experience confinement.

Another interpretation of the phonon modes of the III-V nanostructures appearing in the TO-LO gap was given by Gupta et al. [34] in the case of GaP. According to this interpretation the surface modes of the GaP nanorods (diameter 70nm) dominate the spectra due to the large surface/volume ratio. The frequency of this surface modes is determined by the electric field on the surface [35]. A third explanation could arise from diameter variations of the nanorods, especially periodic ones occurring through the growth process. They could lead to zone folding effects for k-vectors along the axis of the nanorod . The effect on the low dispersive TO phonon would be small but the LO phonons, having the stronger dispersion, could appear with different frequencies.

### 3.2.2. *ZnO nanowires*

Semiconducting ZnO nanowires were prepared in the laboratories of IMEM Parma by thermal evaporation inside a tubular reactor with controlled gas flow. ZnO nanowires of different shapes where obtained with thicknesses between 20 and 200 nm (Fig.2b). Raman measurement presented here (Fig.4) were done under non-resonant conditions ($photonenergy = 2.7eV << E_g = 3.4eV$). Thus the absorbance of the material is very low and higher excitation laser power densities (up to 70 mW/$\mu m^2$) could be applied without the risk of heating effects.

The Raman spectra of ZnO nanowires (¡100 nm diameter) show a clear difference to the ZnO bulk spectra. From the measurements on the sample in Fig. 2b) two different types of spectra were obtained. They are shown in Fig.4a) (type 1) and Fig.4b) (type 2). The shown spectra occurred reproducible at different locations on the sample. Type 1 (Fig. 4a) shows the spectra which result from the central region of this specific sample.

(a)

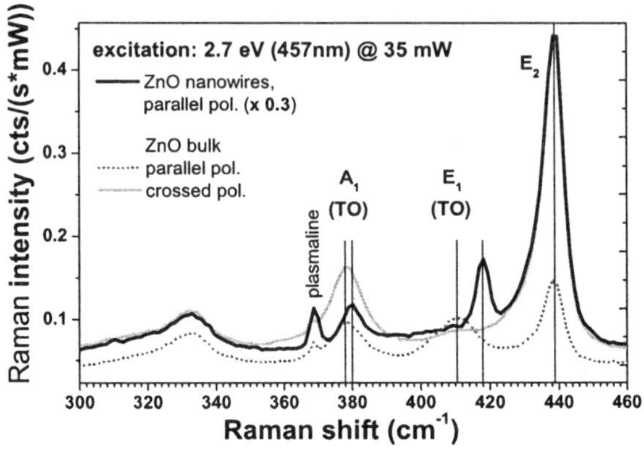


(b)

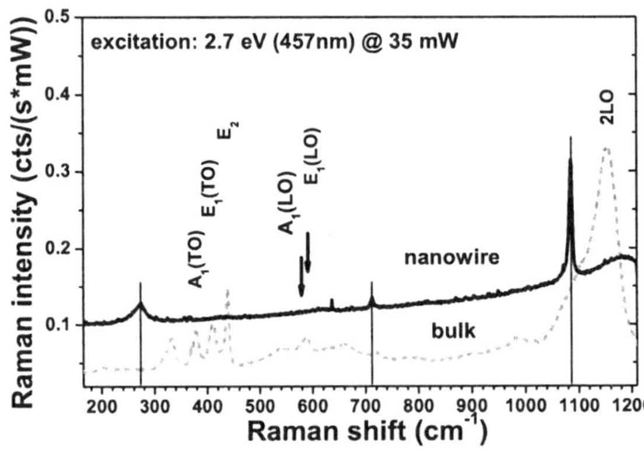


Figure 4. Raman spectra of ZnO nanowires at different positions on the sample (Fig. 2b))in comparison to ZnO bulk spectra. (a) Type 1 spectra appear from the central region of the sample. A new line appears in the gap between the $E_1$(TO) and $E_2$ at approx. the 418 $cm^{-1}$. Moreover the $A_1$(TO) line is shifted by ca. 2 $cm^{-1}$ to the higher wave numbers. (b) Type 2 spectra appearing on the borders of the sample. Only the double phonon lines seem to contribute significantly to the Raman signal. A narrow intense double phonon line appear at ca. 1100 $cm^{-1}$.

Type 2 spectra were measured on the border region of the sample. The two graphs are shown in comparison to ZnO bulk spectra. In the type 1 spectra a new line appears in the gap between the $E_1$(TO) and $E_2$ always at the same frequency. Moreover, the $A_1$(TO) line is shifted to the higher wave numbers. Even though this shift seems to be small it has a significant value of ca. 2 $cm^{-1}$ and can not be a result of possible calibration inaccuracy due to direct comparison of the line energies of other lines ($E_2$ and the double phonon line at 343 $cm^{-1}$). According to the bulk phonon dispersion relation of ZnO [36], the new lines or the line shifts could be the result of folded optical phonon branches in the directions from $\Gamma$ point to the directions A, M or K.

The type 2 spectra (Fig. 4b)) are very different from the previous and the ZnO bulk spectra. Only the double phonon lines seem to contribute significantly to the Raman signal. The appearance of the narrow intensive 2LO line at ca. 1100 $cm^{-1}$ suggest the presence of a phonon selection mechanism. For example because of the small size (confinement) combined with an asymmetric shape of the nano-structure (phonon k-vector selection). The fact that two different types of spectra originate from the same sample, is probably due to the slightly different growth condition in its center and on its borders, resulting in nano-structures of different sizes and types.

The appearance of two types of Raman spectra demonstrate the need for higher spatial resolution. Measuring the spectra from a single nano structure and at the same time its topography would simplify considerably the interpretation of the obtained data.

## 4. Nano Resolution Raman scattering: From Far Field to Near Field

Since the resolving power of optical instruments in the far field (sample-observer distance $>> \lambda$) is in the order of the wavelength (Equ. 5), it is obvious that the optical information on the structure of smaller structures disappears in between. We demonstrate this by showing simulated diffraction patterns (Fig.5) of slits illuminated by a plane wave for different distances from the slits. The exact calculation of the light transmission trough a small aperture request a self consistent solution of the Maxwell equations [37, 38]. But in order to obtain the qualitative behavior of the diffraction the problem can be also treated by means of a simple scalar theory. The simulated diffraction patterns in Fig.5 were obtained according to Huygens' principles assuming every point in the space between the slit borders to be the origin of a spherical wave and summing up the scalar field intensity of this waves at the desired coordinates:

$$E^2 = \int^T \int^A \frac{\exp^{-i(\mathbf{kr}-\omega t)}}{\mathbf{r}} \mathrm{d}a\mathrm{d}t \qquad (6)$$

where A is the area emitting spherical waves (composed of one or two finite size slits), $a$ the position within the slits, $T >> (1/\omega)$ the integration time,

(a)

(b)

(c)

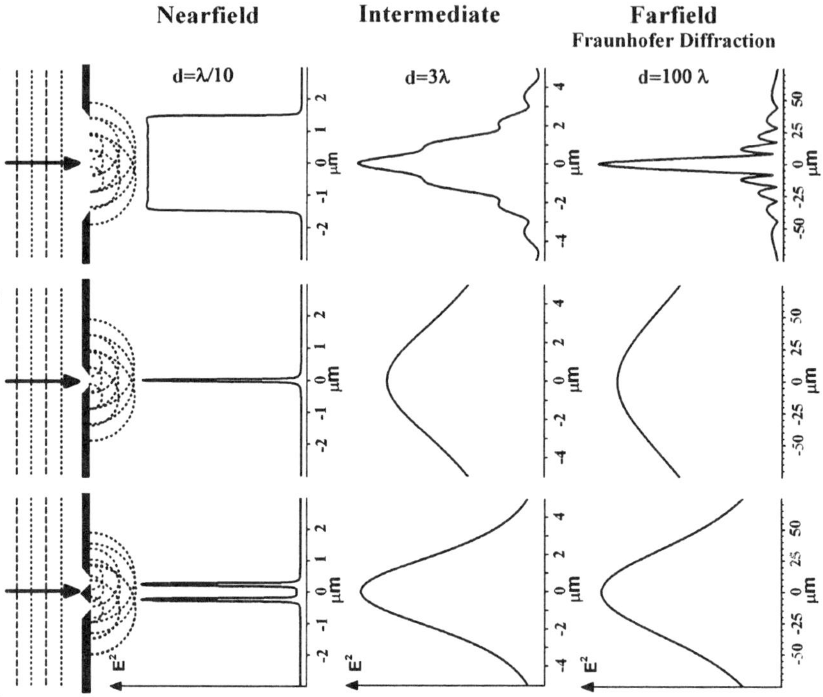


Figure 5. Simulated intensity patterns from infinite slits transmission in the near-, intermediate and far field calculated by summing the secondary spherical waves with the origin in the slits. The slits are illuminated by a plane wave at $\lambda = 500nm$ from the left. The large single slit (on top) has the size of three times the wavelength, the narrow slits width (in the middle part) is $(\lambda/10)$. The double slit (lower part) is consistent out of two $\lambda/10$ slits separated by $\lambda/2$ spacer.

k is the wave vector, $\omega$ the frequency of the light and $\mathbf{r}$ the distance from the origin of a single spherical wave.

The calculated diffraction pattern in Fig.5 shows on the right side ($d = 100\lambda$, far field) the well known Fraunhofer diffraction pattern. For wide slits (w $= 3\lambda$, Fig.5a) the far field diffraction is similar to the well known Airy-function pattern which gives the resolution of Equ. 5. For a smaller slit (in Fig.5b) with slit width d$= \lambda/10$) nearly uniform scattering into the right half space, similar to a simple spherical wave, is observed. The phase difference of the spherical waves originating in the slit is not sufficient to create a diffraction pattern with several maxima. Even in the case of two narrow $\lambda/10$ slits separated by $\lambda/2$ spacer, shown in the lower part of Fig.5, the phase difference (caused by the distance between the slits) is not sufficient to make the two small slits distinguishable in the far field. The far field diffraction patterns of one and two narrow slits are not distinguishable.

However, in the proximity of the slit plane within a wavelength fraction ($\lambda/10$) distance to the slits (Near Field) the diffraction pattern of the single slit (b) and the double slit (c) are clearly distinguishable and display the structure of the object. Thus sub-wavelength structures can be resolved. This sub-wavelength structure information with spatial frequencies larger than $k > \frac{1}{\lambda}$ is carried by the non-propagating near field which does not contribute to the far field pattern [39].

In order to detect the sub-wave length structure of the sample it is necessary to convert the non propagating fields near the surface of the nano-structured sample into the propagating waves. The propagating waves can then be detected by a standard far field optical techniques.

In principle the near field can be transformed into propagating waves by placing any kind of probe in the near field into the vicinity of their source. For high spatial resolution imaging and spectroscopy the probe should be small and could be either a glass fibre or a sharp metallic tip. Excited through the near fields of the sample, the probe irradiates propagating waves. They can be detected in the far field with conventional optical techniques. Due to the rapid distance dependent intensity decrease only the structures in the near field of the probe contribute significantly to the generation of the propagating waves. The spatial resolution corresponds to the size of the probe and its distance from the sample. When scanning a nano structured surface through the near field region parallel to the surface an intensity pattern such as shown in Fig.5 is obtained. This is then called Scanning Near field Optical Microscopy (SNOM). Such instruments utilizing the well developed scanning technique of the scanning probe microscope (AFM, STM) are now commercially available. The basic principle was published already in 1928 by Synge [40], who suggested as a probe a small aperture. The first realization of this concept for the optical spectral range with a glass fiber had to await for more then half a century (D.W. Pohl et al.;1984) [2]. The experimental problem is that the light transmission trough an aperture smaller than the wavelength decreases strongly with its size. Therefore, for intensity reasons the aperture size and consequently the resolution is limited to several tens of nanometers.

The transformation of the non propagating near fields into propagating waves can be obtained also by introducing roughness or small particles onto the surface. Usually thin nonhomogeneous silver films are used for this purpose. The latter is utilized in the so called surface enhanced Raman scattering (SERS) [41]. From the latter studies it was concluded that especially the plasmon resonance (Ag at 3.9eV) enhances the transformation efficiency of near fields into propagating waves since strong dipoles are excited at the plasmon resonance frequency. Thus the Ag particles act as efficient (resonant) antennas which, placed in the near field, radiate into the far field. From these observation it is consequent to use Ag-tips instead of fiber tips to probe the near field. Such configuration have been suggested and realized recently and are named Aperture less Scanning Near field Optical Microscope (A-SNOM). They do not have the problem of the intensity limited resolution as glass fiber tips have.

Even if we are interested here just in a spectroscopic measurement (Raman scattering) with high spatial resolution it is obvious that this optical signal may also be used for image creation of the object features. This could be named a Raman-SNOM. Although this seems to be not feasible because of the general weakness of this higher order optical effect, scanning features can be very useful in order to determine the morphology of the sample and to know where the nano-structures are located on a substrate. However, from signal to noise arguments it will be advantageous to use either a stronger optical effect (transmission or reflection from the laser light). Even better in the case of a conducting sample one can use the metallic tip in a STM mode or if the sample is not conducting operate in an AFM mode. STM and AFM also do not have the disadvantage occurring sometimes in SNOM images through ambiguities in the interpretation of the intensity features. We summarize this section by stating that the tip enhancement not only serves the purpose of high spatial resolution but if used in combination with a scanning microscope technique is also extremely useful in generating an image of the measured object .

## 5. Realizations of Raman Scattering with a tip (SNOM or A-SNOM)

From the forgoing it is clear that with coated optical fibers experimental verifications of optical near field Raman scattering is difficult. Nevertheless it has been already tried since the beginning of the nineties [6], but due to intensity problems fiber applications were limited to samples with large Raman cross sections [5, 42]. On the other hand aperture less techniques based on plasmon related resonant field enhancement, occurring for some metals in the visible spectral range, give a promising outlook to overcome the intensity problem especially severe in Raman spectroscopy.

This plasmon resonance enhancement has been first observed in SERS (Surface Enhanced Raman Scattering) [41, 43] and exploited for very sensi-

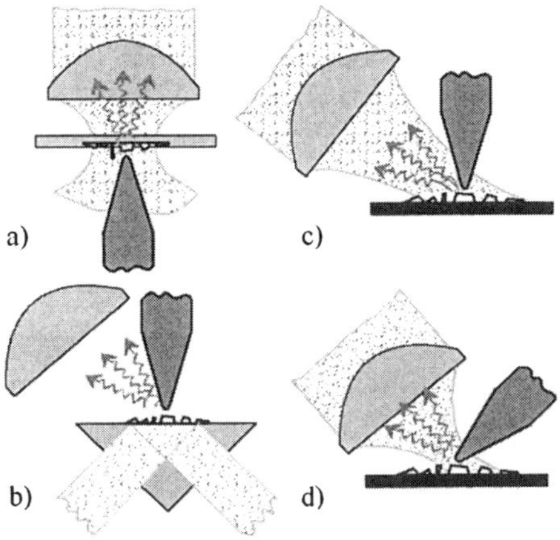


Figure 6. Nano-Raman configurations: a) transmission mode with high NA microscope objective, b) optical tunnelling mode, c) reflection mode, d) optimized reflection mode.

tive molecular identification [11, 44, 45]. Recent progress in computational theory of plasmonic properties of noble metal nano-particles [46] helps to optimize the metal tips with respect to their signal enhancement and the confinement of the near-field. Indeed, very recently aperture less techniques have been combined with a near field Raman scattering set-up for the first time [39, 47]. This concept is also referred to as TERS (Tip Enhanced Raman Spectroscopy) and its successful operation including the tip enhancement effect was shown for carbon nano-tubes [48]. Due to the higher numerical aperture of this configuration those attempts are realized in the transmission mode while only few reflection mode TERS set-ups were exploited [11, 49].

In (Fig. 6) we show possible tip-lens-sample configurations for Raman scattering. A major advantage of the transmission mode (Fig. 6a) is that high numerical apertures are easily achieved using microscope objectives with short working distances and immersion. Thus the light can be efficiently focused and collected. The main disadvantage is that it is limited to samples deposited on transparent substrates.

In the mode represented in (Fig. 6b), the illumination is provided by attenuated total reflection of a laser beam within a prism. Hence, in this configuration tip and sample are illuminated by the evanescent field which is generated at the interface while the illumination far field radiation is reflected into the prism. This configuration has a favorable signal-to-noise ratio as the tip is enhancing only the near-field, whereas in other configurations in Fig. 6 the far field is detected simultaneously with the near field. However, the optical properties of the prism may limit the spectral range of laser excitations or generate additional luminescence background. In analogy to the STM this configuration is usually called optical tunnelling mode.

A reflection mode is needed in order to explore samples on non-transparent substrates (Fig. 6c). In this geometry the tip (including distance control) and the optical set-up illuminating the tip and collecting the scattered light have to be arranged on the same side of the sample. As a consequence only long distance microscope objectives can be used and the numerical aperture available in this configuration is considerably decreased.

The reflection mode can be optimized with respect to the achieved field enhancement because it depends strongly on the illumination geometry [50]. The strongest field enhancement below the tip apex is observed if the polarization of the incident field is polarized along the axis of the tip. Furthermore, in theoretical models the tip is in general described by one or several dipoles. Considering their emission, a 90°-configuration is likewise favorable to efficiently collect the enhanced light (Fig. 6d).

In order to achieve a resolution in the nanometer range the set-up has to be isolated from vibrations. For this reason the whole set-up for Raman spectroscopy (Fig. 1)is placed on a so-called optical table, which damps strongly the perturbation of vibrations from the laboratory surroundings.

While for STM applications only the foremost atoms of the tip play a major role for the resolution (tunnelling current), whereas the conical part of the tip is not of importance, in optical applications the mesoscopic structure of the tip (i.e. radius of curvature at the apex, angle of the cone) plays the central role. Furthermore, due to the relation of the enhancement with its dielectric properties the material of the tip has to be chosen according to the spectral region studied. For the use in the spectral range from 350 nm to 515 nm silver has shown to have a high enhancement factor. Hence, the probes consist of sharp silver tips with an apex radius clearly below 100 nm. To produce silver nano-tips an electrochemical-dynamical etching procedure was developed. It consists in electrochemical etching of a $0.25mm$ diameter silver wire using a highly conductive aqueous solution (high solubility inorganic salt dissolved in water) and applying a voltage $(1 - 5V)$ between the wire (anode) and a graphite counter electrode (cathode). During the etching the wire is retracted from the solution by a small pitch electric motor. This procedure allows to get very sharp tips (apex diameter $< 100nm$) and of a shape suitable for A-SNOM applications.

The xy-scanning and the sample positioning are achieved by translation stages with a sub-nanometer resolution provided by the use of piezo-electric elements. The probe-sample distance can be controlled either by measuring the tunneling current in the STM mode (limited to conducting sample-substrate systems) or determining the forces between probe and sample in the AFM [8] mode. In the AFM mode a tuning fork sensor is used in the shear-force feedback mode. The shear-force mode is a non-contact mode which bases on the dependence of the damping of the oscillation on the distance between probe and sample. While for a resolution in z direction a very high Q-factor is preferable, a high Q-factors limits the scanning speed due to its slower adaption to the changing conditions [51]. To allow high Q-factors and a high scanning speed, it was proposed to use the shift of the resonance frequency as a feedback signal [52]. Both STM and AFM mode can operate in a constant signal mode, where the sample-probe distance is kept constant. This is important because the intensity of the optical signal is strongly dependent on the sample-probe distance and should be not falsified by distance variations.

The optical part of set-up is designed in a reflection geometry allowing us to work with non-transparent samples. Illumination and detection of the Raman signal are obtained with a microscope objective with a long working distance (4 mm) and a comparably high numerical aperture (0.75). To collect the enhanced Raman signal more efficiently both, objective and tip, are tilted (Fig. 6d).

## 6. Summary

Measurements to test the developed Nano Raman equipment were performed on Rhodamine 6G adsorbed on a silver substrate. To prepare the samples Rhodamine 6G was solved in alcohol and applied to the silver substrate. The topography of the prepared samples(Fig. 7a) was determined by the STM with the same silver tip which was used for the spectroscopic measurements. Obviously by this method of sample preparation it was not possible to obtain a monolayer of molecules on the surface instead Rhodamine 6G agglomerates formed on the surface (the distinct features in the STM image). Although, in some cases it was possible to identify agglomerates on the surface, the non-conductivity of the Rhodamine 6G agglomerates turned out to be a severe obstacle for STM measurements.

In order to measure Nano Raman spectra of the adsorbed molecules tip and sample were brought into tunnelling contact (upper spectrum Fig. 7 b) in a region of the sample with a low density of Rhodamine 6G agglomerates. The sample was excited with the 488 nm of the $Ar^+$ laser (2.54 eV) with a power of 0.8 mW. To determine the far field background the sample was retracted by about 0.5 $\mu$m from the tip measuring the lower spectrum (Fig.

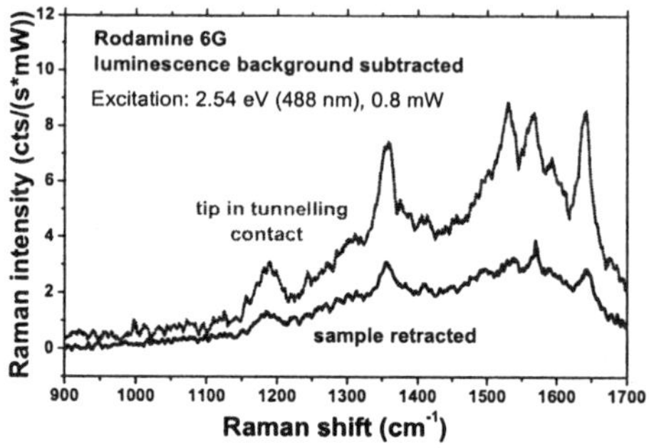


(a) (b)

Figure 7. Nano Raman test measurements on Rhodamine 6G. On the left: "typical" STM measurements of Rhodamine 6G agglomerates on silver surface. On the right: nano Raman measurements with tip in tunnelling (upper spectra), lower spectra is taken after retracting the sample by ca. 0.5 $\mu$m.

7b). Both spectra are shown without the luminescence background. Even though the primary aim of the measurement was to test the spectroscopic mode we observed a good agreement with near field Raman spectra reported for Rhodamine 6G [45]. Nevertheless, we do not consider the reported spectrum very reliable, because the sample was retracted from the tip, while focus of the exciting laser remained on the apex of the tip, so that Rhodamine 6G that might be attached to the tip contribute to the far field background. In order to increase the reliability of measured near field spectra the tip has to be retracted from the sample and the focus of the exciting laser, and repositioned with a precision in the sub-nanometer range.

Although some practical improvements seem to be necessary, utilizing the presented experimental set-up, designed according to the principles introduced above, we are confident to be able to perform reliable spectroscopic measurements with spatial resolution in the nanometer range.

## Acknowledgements

We thank Antonio Cricenti for the excellent organization of the Epioptics-9 conference and the possibility to participate to this exceptional meeting. We would also like to express our special acknowledgements to Nico Lovergine and his coworkers for providing GaAs nanorod samples of astonishing quality which enabled us to perform an important part of the measurements presented here.

**References**


1. G. Binnig, H. Rohrer, C. Gerber, and E. Weibel. *Appl. Phys. Lett.*, **40**:178–180, 1982.
2. D.W. Pohl, W. Denk, and M. Lanz. *Appl. Phys. Lett.*, **44**:651, 1984.
3. B. Hecht, B. Sick, and U.P. Wild et al. *J. Chem. Phys.*, **112**:7761–7774, 2000.
4. L. Novotny and C. Hafner. *Phys. Rev. E*, **50**:4094, 1994.
5. M. Goetz, D. Drews, D.R.T. Zahn, and R. Wannemacher. *J. of Luminescence*, **76**:306, 1998.
6. C.L. Jahncke, M.A. Paesler, and H.D. Hallen. *Appl. Phys. Lett.*, **67**:2484, 1995.
7. F. Zenhausern, M.P.O'Boyle, and H.K. Wickramasinghe. *Appl. Phys. Lett.*, **65**:1623, 1994.
8. G. Binnig, C.F. Quate, and C. Gerber. *Phys. Rev. Lett.*, **56**:930–933, 1986.
9. N. Hayazawa, Y. Inouye, Z. Sekkat, and S. Kawata. *Chem. Phys. Lett.*, **335**:369, 2001.
10. Mark S. Anderson. *Appl. Phys. Lett.*, **76**:3130, 2000.
11. B. Pettinger, B. Ren, G. Picardi, R. Schuster, and G. Ertl. *Phys. Rev. Lett.*, **92**:096101, 2004.
12. T. Ruf. *PhononRaman Scattering in Semiconductors, Quantum Wells and Superlattices, Springer Tracts in Modern Physics.* Sringer-Verlag, Berlin, Heidelberg, 1998.
13. R. Loudon. *Proc. Royal Soc.*, **A275**:218, 1963.
14. W. Hayes and R. Loudon. *Light Scattering in Solids.* J. Wiley and Sons, New York, 1978.
15. W. Richter. *Springer Tracts in Modern Physics **Vol. 78**, Resonant Raman Scattering in Semiconductors, ed. by G. Höhler.* Springer, Berlin, Heidelberg, New York, 1976.
16. M. Cardona. *Topics in Applied Physics **Vol. 50**, Light Scattering in Solids II, ed. by M. Cardona and G. Güntherodt.* Springer, Berlin, Heidelberg, New York, 1982.
17. E. Speiser, K. Fleischer, and W. Richter. *Epioptics-8, Proceedings of the 33rd Course of the International School of Solid State Physics, Page 92, ed. by Antonio Cricenti.* World Scientific, 2006.
18. M. Cardona. *Light Scattering in Solids V. Superlattices and Other Microstructures, ed. by M. Cardona and G. Güntherodt.* Springer-Verlag Berlin and Heidelberg GmbH & Co. K, 1989.
19. N. Esser and J. Geurts. *Optical Characterization of Epitaxial Semiconductor Layers, Ed.: G. Bauer, W. Richter.* Springer, Berlin, Heidelberg, New York, 1996.
20. A. Pinczuk and E. Burstein. *Topics in Applied Physics **Vol. 8**, Light Scattering in Solids, ed. by M. Cardona and G. Güntherodt.* Springer, Berlin, Heidelberg, New York, 1975.
21. L. Novotny and B. Hecht. *Principles of Nano-Optics.* Cambridge University Press, 2006.
22. M.V. Klein. *Optics.* J. Wiley and Sons, New York, 1970.
23. R.H. Webb. *Rep. Prog. Phys.*, **59**:427–471, 1996.
24. G.S. Duesberg, W.J. Blau, H.J. Byrne, J. Muster, M. Burghard, and S. Roth.

25. G. S. Duesberg, I. Loa, M. Burghard, K. Syassen, and S. Roth. *Phys. Rev. Lett.*, **85**:5436, 2000.
26. A. Jorio, A. G. Souza Filho, G. Dresselhaus, M. S. Dresselhaus, A. K. Swan, M. S. nl, B. B. Goldberg, M. A. Pimenta, J. H. Hafner, C. M. Lieber, and R. Saito. *Phys. Rev.*, **65**:155412, 2002.
27. A. Jorio, M.A. Pimental, A.G. Souza Filho, R. Saito, G. Dresselhaus, and M.S. Dresselhaus. *New J. of Phys.*, **5**:139.1, 2003.
28. J. Maultzsch, S. Reich, and C. Thomsen. *Phys. Rev.*, **65**:233402–1, 2002.
29. P. Paiano, P. Prete, N. Lovergine, and A. M. Mancini. *J. Appl. Phys.*, **100**:094305, 2006.
30. R.S. Wagner and W.C. Ellis. *Appl. Phys. Lett.*, **4**:89–90, 1964.
31. P. Paiano, P. Prete, E. Speiser, N. Lovergine, W. Richter, L. Tapfer, and A.M. Mancini. *J. of Crys. Growth*, **298**:620, 2007.
32. E. Speiser, T. Schmidtling, K. Fleischer, N. Esser, and W. Richter. *phys. stat. sol (c)*, **0**:2949–2955, 2003.
33. J. Menendez and M. Cardona. *Phys. Rev.*, **B 29**:2051, 1984.
34. Rajeev Gupta, Q. Xiong, G. D. Mahan, and P. C. Eklund. *Nano Letters*, **3**:1745, 2003.
35. Bo E. Sernelius. *Surface Modes in Physics*. Wiley-VCH, New York, 2001.
36. J. Serrano, F. Widulle, A. H. Romero, A. Rubio, R. Lauck, and M. Cardona. *phys. stat. sol. (b)*, **235**:260, 2003.
37. H. Bethe. *Phys. Rev.*, **66**:163, 1944.
38. C.J. Bouwkamp. *Rep. Prog. Phys.*, **17**:35, 1954.
39. A. Hartschuh, E.J. Snchez, X.S. Xie, and L Novotny. *Phys. Rev. Lett.*, **90**:095503, 2003.
40. E.H. Synge. *Phil. Mag.*, **6**:356, 1928.
41. M. Moskovits. *Rev. Mod. Phys.*, **57**:783, 1985.
42. C.L. Jahncke, H.D. Hallen, and M.A. Paesler. *J. of Raman Spec.*, **27**:579, 1996.
43. M. Fleischmann, P.J. Hendra, and A.J. McQuillan. *Chem. Phys. Lett.*, **26**:163, 1974.
44. T. Ichimura, N. Hayazawa, M. Hashimoto, Y. Inouye, and S. Kawata. *Phys. Rev. Lett.*, **92**:220801, 2004.
45. N. Hayazawa, Y. Inouye, Z. Sekkat, and S. Kawata. *Chem. Phys. Lett.*, **335**:369, 2001.
46. Cecilia Noguez. *Chem. Phys. Lett.*, **27**:1204, 2005.
47. N. Hayazawa, T. Yano, H. Watanabe, Y. Inouye, and S. Kawata. *Chem. Phys. Lett.*, **376**:174, 2003.
48. Achim Hartschuh, Erik J. Sanchez, X. Sunney Xie, and Lukas Novotny. *Phys. Rev. Lett.*, **90**:095503–1, 2003.
49. Y. Saito, M. Motohashi, and N. Hayazawa. *Appl. Phys. Lett.*, **88**:143109, 2006.
50. L Novotny, R.X. Bian, and X.S. Xie. *Phys. Rev. Lett.*, **79**:4, 1997.
51. Khaled Karrai and Robert D. Grober. *Appl. Phys. Lett.*, **66**:1842, 1995.
52. T.R. Albrecht, P. Grttner, D. Horne, and D. Rugar. *J. Appl. Phys.*, **68**:668, 1989.

# INVESTIGATION OF COMPOSITIONAL DISORDER IN $GaAs_{1-x}N_x$:H

R. TROTTA, M. FELICI, F. MASIA, A. POLIMENI, A. MIRIAMETRO, AND M. CAPIZZI
*Dipartimento di fisica , Università di Roma "La Sapienza", P.le A. Moro 2 Roma, 00185, Italy*

P. J. KLAR, AND W. STOLZ
*Department of Physics and Material Sciences Centre, Phillips-University, Renthof 5 Marburg, D-35032, Germany*

Compositional disorder is investigated by means of photoluminescence (PL) and PL excitation (PLE) measurements in as-grown and hydrogen-irradiated $GaAs_{1-x}N_x$ samples ($x \leq 0.21\%$). The dependence of the linewidth of the PLE free-exciton (FE) band on N concentration agrees well with that predicted by a theoretical model developed for a purely random alloy. We also find that H irradiation and ensuing nitrogen passivation reduce significantly the broadening of the FE band. This result is consistent with an H induced removal of the static disorder caused by N. Finally, an analysis of the dependence of the Stokes Shift on the FE linewidth shows that free carriers are thermalized even at low temperature, another indication of a low degree of disorder in the investigated samples.

## 1. Introduction

The striking effects of nitrogen incorporation on the electronic properties of $GaAs_{1-x}N_x$ have been studied in great detail in the past decade[1]. In particular, the giant bowing of the energy gap and an uncommonly large and abrupt increase in the electron effective mass with the increasing N concentration have been the object of many reports, both theoretical[2] and experimental[3].

The response to irradiation with atomic hydrogen is another peculiarity of the $GaAs_{1-x}N_x$ alloy. Recently, it was shown that hydrogen irradiation counteracts all electronics and structural effects produced by N introduction in the host lattice of dilute nitrides, and finally transmutes hydrogenated $GaAs_{1-x}N_x$ into "pure" GaAs[4]. This intriguing behavior is explained by the formation of N-H complexes that neutralize the perturbation exerted by N atoms on the host matrix[5]. Although a fine tuning of the energy gap and of the electron effective mass by post-growth hydrogenation might have a wide range of applications, further studies are required before fabricating devices or nanostructures based on hydrogenated $GaAs_{1-x}N_x$. To this respect, an analysis of the compositional and structural disorder of $GaAs_{1-x}N_x$ samples, and of its change upon H irradiation is an issue of special interest.

In this context, we report here an optical investigation of the compositional disorder in $GaAs_{1-x}N_x$ epilayers at N concentration less than or equal to 0.21%, where a detailed study of N atoms distribution is still lacking, both in as-grown and hydrogenated samples. By photoluminescence (PL) and PL excitation (PLE) measurements we studied the dependence on N concentration of the linewidth of the PLE free exciton band. This dependence is reproduced very well by a statistical model, which is valid only in the limit of a purely random alloy. Moreover and quite surprising, the FE linewidth decreases with increasing hydrogen dose, thus providing further experimental support to a restoration of the lattice potential order upon hydrogenation. Finally, the relation between the Stokes Shift and the FE linewidth is compared with a model describing carrier relaxation in the presence of alloy disorder. Such comparison shows that free-excitons are thermalized even at low temperature, thus confirming the low degree of disorder in these ternary alloys at low $x$.

## 2. Discussion and results

We studied three $GaAs_{1-x}N_x$/GaAs epitaxial layers (having $x$=0.043%, 0.095%, and 0.21%) grown by metalorganic vapor phase epitaxy and post growth hydrogenated at increasing H doses. Details about hydrogenation and optical detection have been given elsewhere[6]. PL and PLE spectra of the as-grown samples as well as of two hydrogenated samples are shown in Fig. 1 (*a*). Labels in the PL spectra indicate the $GaAs_{1-x}N_x$ free-exciton band (downward pointing arrows), the free electron to carbon acceptor band (e,C) and bands from different N cluster, NC. In PLE upward pointing arrows mark the $GaAs_{1-x}N_x$ FE band. The GaAs FE peak at 1.515 eV is observed together with features at lower energy likely due to N cluster states resonant with the continuum of the above-gap states of $GaAs_{1-x}N_x$. The spectra of the hydrogenated samples illustrate the progressive passivation of all nitrogen induced effects by means of H irradiation and that the FE linewidth reduces sizably with increasing hydrogen dose.

As exemplified in Fig. 1 (*b*) for an untreated sample with $x$=0.095%, we find that both a light- (*lh*) and a heavy-hole (*hh*) component at low and high energy, respectively, contribute to the FE band in the PLE spectrum. Indeed, the **k**=0 valence band degeneracy is removed by tensile strain due to the lattice mismatch between the $GaAs_{1-x}N_x$ epitaxial layer and the GaAs substrate. A detailed study of the evolution of this splitting with increasing N concentration and H dose performed by means of polarization-dependent PLE measurements will be reported elsewhere[7].

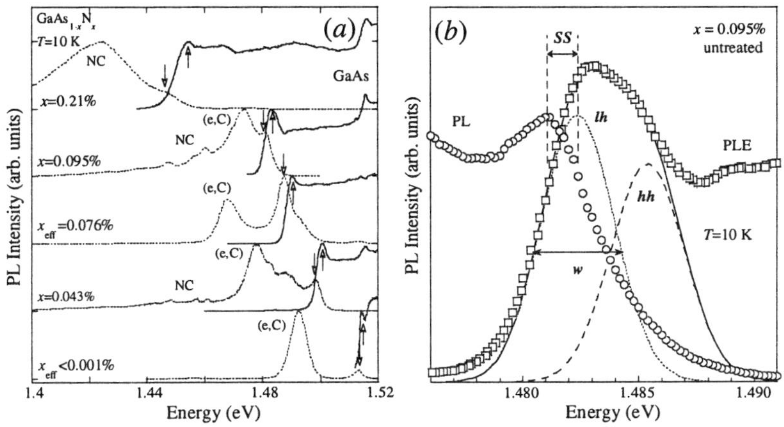


Figure 1 (*a*): Peak-normalized photoluminescence (PL, dotted line) and PL excitation (PLE, continuous line) spectra at $T$=10 K of a selection of the $GaAs_{1-x}N_x$ samples studied in this work. For the hydrogenated samples, the effective N concentration $x_{eff}$ is also specified (it was estimated by comparing the energy position in PLE of the FE peak with a calibration curve). Upward and downward arrows mark the free-exciton energy for PLE and PL spectra, respectively. (e,C) indicates the free-electron to neutral-carbon recombination, and NC indicates carrier recombination from nitrogen complexes. Figure 1 (*b*): Comparison between PL (circle) and PLE (squares) spectra at $T$=10 K of an untreated $x$=0.095% sample. The result of a fit of the free-exciton PLE band using two Gaussians is also shown (continuous line).

We now compare the experimental linewidth $w$ of the light-hole component of the PLE excitonic band (extracted by fitting the experimental data with two Gaussians like those shown in Fig. 1 (*b*)) with its theoretical estimate obtained following Ref. 8. In a generic random alloy $AB_{1-x}C_x$, the free-exciton band has a Gaussian lineshape with a full-width at half-maximum $w$, whose dependence on $x$ is given by[8]

$$w = \gamma\left( 2\frac{dE_{FE}}{dx}\sqrt{\frac{2\ln 2(1-x)xV_c}{(4\pi/3)R_{ex}^3}} \right) \qquad (1)$$

According to this model, in a random alloy with low structural disorder, the broadening of the excitonic band is entirely due to the potential fluctuations of N concentration that free excitons sample within a characteristic length given by Bohr radius, $R_{ex}$. $V_c=a^3/4$ is the smallest volume over which an alloy fluctuation can occur, and the lattice constant $a$ is kept fixed to the value of GaAs (5.653 Å). $\gamma$ is the only fitting parameter, introduced because of a certain degree of arbitrariness in the definition of $V_c$ and $R_{ex}$. In order to calculate the FE broadening from Eq. (1), we experimentally estimate the exciton Bohr radius directly from the diamagnetic shift of the photoluminescence FE peak[9]. The dependence of the energy gap on N concentration $E_{FE}(x) = E^0_{FE}(x)\text{-}16.2{\cdot}x^{0.9}$, in eV, has been obtained by fitting a power-law to the PLE values of $E_{FE}$ and agrees well with the theoretical prediction[10] of a power exponent ranging between 0.66 and 0.89.

The values of the full-width at half-maximum of the FE band as obtained from Eq. (1) are compared in Fig. 2 (*a*) with the experimental values of *w*. The agreement between experiment (full symbols) and theory (open squares) is quite good, both for the as-grown (full circles) and the hydrogenated (full triangles) samples.

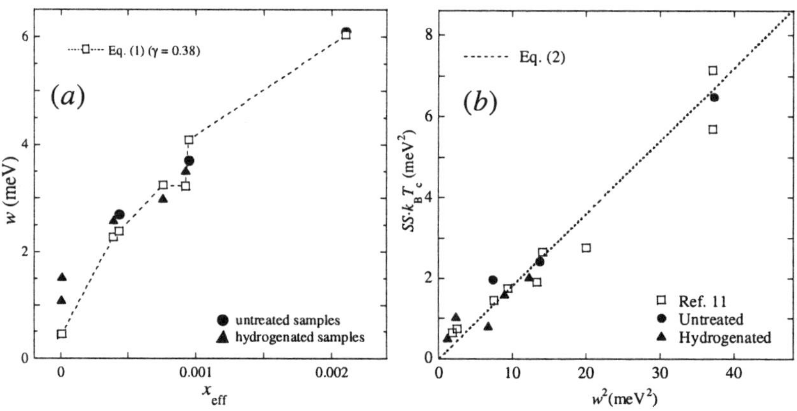


Figure 2 (*a*): Dependence of the experimental free-exciton linewidth *w* (full symbols) on the effective nitrogen concentration $x_{eff}$. Full circles refer to as-grown samples, whereas full triangles refer to the hydrogenated epilayers. The theoretical predictions of Eq. (1) are also shown in the figure (open squares; the dashed line is a guide for the eye). Figure 2 (*b*): Dependence of the product of the Stokes Shift *SS* and the carrier thermal energy $k_BT_c$ on the linewidth squared $w^2$. *w* values as derived from PLE and $k_BT_c$ has been estimated from the ratio between PL and PLE spectra, as discussed in the text. Full circles (triangles) refer to as-grown (hydrogenated) samples. The dotted line gives the behavior described by Eq. (4). Data from Ref. 11 (empty squares) are also shown for comparison.

The value of $\gamma$ that best fits the experimental data shown in Fig. 2 (*a*) is 0.38, very close to the value of 0.41 estimated in Ref. 12. Finally, the small, abrupt "jump" occurring in the FE linewidth at about $x$=0.1% is the signature of a shrinking of $R_{ex}$. This feature was well established in several previous papers and is due to a sudden change in the electron effective mass value[9]. The agreement we found between the experimental observations and the predictions of Eq. (1) provides strong evidence that nitrogen atoms are randomly distributed in $GaAs_{1-x}N_x$ for $x \leq 0.21\%$, a precondition for the applicability of the model of Ref. 8, and that the hydrogenation process does not affect severely the randomicity of the N atom distribution. Most interestingly, the hydrogenation *decreases* the FE linewidth, a sign of a sizably diminished disorder in the sample. This result is fully consistent with previous experimental[13] findings and theoretical calculations[14] in $GaAs_{1-x}N_x$ alloys.

As shown in Fig. 1 (*b*) we notice that the FE band is red-shifted in PL with respect to its energy in PLE by an amount labeled *SS* (Stokes Shift). This energy difference is due to the fact that PL spectra are dominated by relaxation and thermalization of photoexcited carriers. In semiconductor alloys, the interplay between compositional disorder and carrier thermalization strongly influences the establishment of equilibrium conditions for free-excitons. These processes are investigated here by an analysis of the dependence of the Stokes Shift on the degree of compositional disorder, as measured by the FE linewidth. Among the large number of physical phenomena involved in the relaxation process, two main regimes were previously identified. In highly disordered systems, excitons relax toward and are trapped by local energy minima due to impurities or lattice defects. In this case[15], a linear relationship between *SS* and the FE linewidth is found. In systems with a low degree of disorder the broadening of the excitonic DOS is comparable with the carrier thermal energy. In this case[16], the PL intensity of the FE band is roughly proportional to the excitonic DOS, properly weighted with a Boltzmann distribution function. In quasi-equilibrium conditions, the effective carrier temperature, $T_c$, (which is greater than or equal to that of the lattice) can be determined by an exponential fit of the ratio between the PL and absorption spectra (or PLE, with a good approximation) of the free-exciton.

In this model, whenever the broadening of the excitonic band can be described by a Gaussian function, a simple relationship between *SS*, *w*, and $T_c$ holds[16],

$$SS \cdot k_B T_c = 0.18 \cdot w^2 \ . \qquad (2)$$

In Fig. 2 (*b*) the product $SS \cdot k_B T_c$ is plotted as a function of $w^2$ for all samples (full symbols) together with the behavior predicted by Eq. (2) (dashed line). The quite good agreement between theory and experiment confirms the fundamental role played by thermalization in our samples. Data taken from Ref. 11 and relative to $In_xGa_{1-x}As/GaAs$ quantum wells are also shown in Fig. 2 (*b*) as an example of a random, virtual crystal approximation-like alloy, which behaves according to the thermalization model.

## 3. Conclusions

In summary, we have shown that a fully statistical model[8] for random alloys well describes the dependence of the FE linewidth on nitrogen concentration in $GaAs_{1-x}N_x$, both as-grown and irradiated with hydrogen. Therefore, as-grown $GaAs_{1-x}N_x$ can be treated as a random alloy with a very low structural disorder, at least in the range of N concentration less than or equal to 0.21%. In agreement with previous structural measurements[13], crystalline order is largely recovered upon irradiation with atomic hydrogen as demonstrated by the sizable decrease in the FE band linewidth of hydrogenated samples. Finally, the low degree of disorder in our samples is confirmed by unambiguous evidence that the free-exciton distribution after relaxation is ruled primarily by thermalization.

## References

1. For a review see *"Dilute Nitride Semiconductors: Physics and Technology"*, edited by M. Henini (Elsevier, Oxford, U.K., 2004).
2. A. Lindsay, and E. P. O'Reilly, *Phys. Rev. Lett.* **93**, 196402 (2004).
3. F. Masia *et al.*, *Phys. Rev. B* **73**, 073201 (2006).
4. A. Polimeni *et al.*, *Phys. Rev. B* **69**, 041201 (2006).
5. Mao-Hua Du, S. Limpijumnong, and S.B. Zhang, *Phys. Rev. B* **72**, 073202 (2005).
6. M. Grassi Alessi *et al.*, *Phys. Rev. B* **61**, 10985 (2000).
7. R. Trotta *et al.*, unpublished.
8. R. T. Senger and K. K. Bajaj, *J. Appl. Phys.* **94**, 7505 (2003).
9. F. Masia, *et al.*, *Appl. Phys. Lett.* **82**, 4474 (2003).
10. P. R. C. Kent and Alex Zunger, *Phys. Rev. B* **64**, 115208 (2001).
11. A. Polimeni *et al.*, *Phys. Rev. B* **54**, 16389 (1996).
12. S. M. Lee and K. K. Bajaj, *J. Appl. Phys.* **73**, 1789 (1993).
13. G. Ciatto *et al.*, *Phys. Rev. B* **72**, 085322 (2005).
14. S. Sanna and V. Fiorentini, *Phys. Rev.* B **69**, 125208 (2004).
15. F.Yang *et al.*, *Phys. Rev. Lett.* **70**, 323 (1993).
16. M. Gurioli *et al.*, *Phys. Rev. B* **50**, 11817 (1994).

# VIBRATIONAL PROPERTIES AND THE MINIBAND EFFECT IN InGaAs/InP SUPERLATTICES

A. D. RODRIGUES and J. C. GALZERANI

*Departamento de Física, Universidade Federal de São Carlos,CP 676, 13565-905 São Carlos, Brazil*

YU. A. PUSEP

*Instituto de Física, Universidade de São Paulo,13560-970 São Carlos, Brazil*

D. M. CORNET, D. COMEDI AND R. R. LA PIERRE

*Centre for Emerging Device Technologies, McMaster University, Hamilton, Ontario, Canada L8S 4L7*

InGaAs/InP:Si superlattices (SL's) of several periods were investigated using polarized Raman spectroscopy. In the long period SL, the translational symmetry was violated and the Raman selection rules were not observed. However, with the decrease of the barrier thicknesses, these rules became evident and considerable blue shifts of the LO modes are noticed, as a manifestation of the presence of coupled plasmon–LO phonon modes propagating normal to the layers in these short period SL's. We here show that the observed effects can be attributed to the formation of a miniband energy structure. The good agreement between the frequencies of the measured InP-like coupled modes with the calculated ones confirms that the blue shifts of the longitudinal modes were observed due to the miniband effect. Further investigations led to the conclusion that the spatial coherence length of the lattice vibrations for the long period SL is of the order of the size of the crystal unit cell; meanwhile, this parameter increases considerably for the short period SL's, also in agreement with the emergence of the selection rules for the coupled vibrations in the later structure.

## 1. INTRODUCTION

The semiconductor superlattices (SL's) — nanostructures consisting of a set of periodic layers made of two different materials — have received special attention since the progress in the growth techniques, specially the Molecular Beam Epitaxy (MBE), allowed to improve the accuracy in the thicknesses and in the control of the composition of theses materials. Raman scattering spectroscopy has been presented as a powerful tool to study this kind of structure, and specially of superlattices made of disordered semiconductors, when the structural and electrical characterizations are required.

Compositional disorder can be deliberately produced in a semiconductor material when, during the growth, conveniently chosen (according to their atomic number and valence) atomic elements are incorporated in the material, thus contributing with an excess of carriers. During this doping process, the excess free carriers can interact, thus originating a collective excitation of electrons, whose resonance frequency ($\omega_p$) depends on the carriers concentration; this quantum of energy is called *plasmon.*

This oscillation is allowed to interact with the electric field associated to the longitudinal optical (LO) mode of vibration of the lattice ions, producing a coupled LO–plasmon mode, whose energy is blue shifted in comparison to the pure LO mode. When the proper conditions occur, the Raman spectrum of a highly doped semiconductor no longer shows the pure LO mode, that is replaced by the $\omega_p$ dependent coupled LO–plasmon mode.

The plasma frequency of the electrons propagating in the growth direction $z$ of a SL is given by $\omega_{pz} = (4\pi e^2 n/\varepsilon_\infty m_z)^{\frac{1}{2}}$, where $\varepsilon_\infty$ is the SL high frequency dielectric constant, $n$ is the carrier concentration and $m_z$ its effective mass in this direction (calculated using the envelope function approximation [1]). In order to calculate $\omega_{pz}$ for a SL such as InGaAs/InP which is doped with Si, we can take the band parameters given in Ref. 2. With this value, it is possible to determine the frequencies of the coupled LO–plasmon phonon modes, calculated as zeros of the diagonal component of the SL dielectric function tensor, in the long wavelength approximation, as in Ref. 3.

$$\epsilon_z(\omega) = \frac{d_1 + d_2}{d_1/\epsilon_1(\omega) + d_2/\epsilon_2(\omega)} - \frac{\omega_{pz}^2}{\omega^2}. \tag{1}$$

Where $\epsilon_1(\omega)$ is the dielectric function of the InGaAs layers, given by

$$\epsilon_1(\omega) = \epsilon_{\infty 1}\left(\frac{\omega_{L1}^2 - \omega^2}{\omega_{T1}^2 - \omega^2} + \frac{\omega_{L2}^2 - \omega^2}{\omega_{T2}^2 - \omega^2}\right) \tag{2}$$

and $\epsilon_2(\omega)$ is the dielectric function of the InP layers determined as

$$\epsilon_2(\omega) = \epsilon_{\infty 2}\left(\frac{\omega_{L3}^2 - \omega^2}{\omega_{T3}^2 - \omega^2}\right). \tag{3}$$

Here $d_1$ and $d_2$ are the thicknesses of the layers, $\epsilon_{\infty 1}$ and $\epsilon_{\infty 2}$ are the dielectric functions of the bulk constituting materials; $\omega_{Ln}$ and $\omega_{Tn}$ are the frequencies of the longitudinal and transversal optical phonons of InAs (when $n$=1), GaAs ($n$=2) and InP ($n$=3). The region where the collective excitation (phonon or plasmon) is localized, is determined by a coherence length ($L_c$). According to Richter et al. [4], this parameter can be obtained from the expression for the Raman intensity of the corresponding collective excitation

$$I(\omega) \sim \int \exp\left[-\frac{(q-q_0)^2 L_c^2}{4}\right] \frac{dq}{[\omega - \omega(q)]^2 + (\Gamma/2)^2} \tag{4}$$

in which $q_0$ is the excitation wave vector (or the screening wave vector) and $\omega(q)$ and $\Gamma$ are, respectively, the dispersion and the damping constant of the collective excitation involved in the process.

This localization length is used to spatially characterize the fluctuations of the crystalline potential, in order to obtain informations about the structural properties of the material. More specifically, in InGaAs/InP SL's doped with Si, the coherence length of the coupled LO–plasmon mode can furnish information about the amplitude of this kind of oscillation, that is intimately connected with the dispersions of the energetic levels of the material, as we will be seeing in this work. Besides, we will show how the detailed analysis of the Raman spectra can furnish informations about the existence of the energy minibands and, consequently, how the occurrence of these minibands can influence the dynamical properties of this system.

## 2. EXPERIMENTAL DETAILS

$(In_{0,53}Ga_{0,47}As)_m(InP)_m$ superlattices (SL's) grown on a semi-isolating (001) InP substrate by MBE were analyzed. Both short period ($m$ = 6, 7, 8, 10, 15; repeated 30 times) and long period ($m$ = 68; repeated 20 times) SL's were studied. The SL's barriers were doped with Si (density 5.0 x $10^{17}$ $cm^{-3}$). Polarized Raman spectra of these samples were obtained in the backscattering configurations $z(x',y')\bar{z}$ and $z(x',x')\bar{z}$ where $x'$, $y'$, $z$ denote the crystal directions [110], [$\bar{1}$10] and [001] respectively. The 514,5nm line of an $Ar^+$ laser was used in order to excite the spectra, that were analyzed by a Jobin Yvon T-64000 triple-grating monochromator equipped with a $LN_2$-cooled CCD detector. All the experiments were carried at 10K.

## 3. DISCUSSIONS AND RESULTS

According to the SL's selection rules, the longitudinal optic (LO) mode is active only in the $z(x',y')\bar{z}$ cross-polarized configuration. No optic mode is allowed in the parallel-polarized ($z(x',x')\bar{z}$ configuration). The analysis of Fig. 1 for the sample with $m = 68$ shows the following features: $LO_1$ (InAs) and $LO_2$ (GaAs) corresponding to the InGaAs wells (allowed); $TO_1$ (InAs), $TO_2$ (GaAs) (forbidden); $LO_3$ (InP) (allowed) and $TO_3$ (InP) (forbidden) and $IF_1$ (GaAs interface) (activated by disorder). We observe that the spectra for both the configurations, $z(x', y')\overline{z}$ and $z(x', x')\overline{z}$ are quite similar; moreover, it is quite noticeable the appearance of the forbidden $LO_3$ showing that no selection rules are obeyed for this sample — probably due to the violation of the translational symmetry in the materials constituting the SL. Fig. 2 presents details of the InP lattice vibrations in the

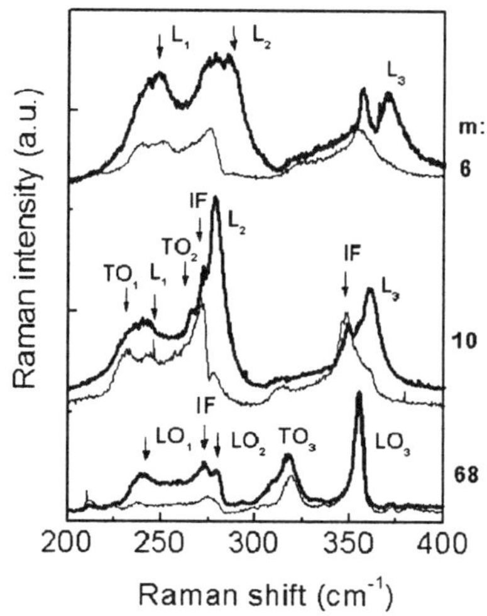


Fig. 1. Raman spectra of the $(InGaAs)_m(InP)_m$ superlattices with different periods. Thin and thick lines correspond to $z(x', x')\overline{z}$ and $z(x', y')\overline{z}$ polarizations respectively. The experiments were accomplished at 10K.

optical range of the spectra. The theoretical analysis of the $LO_3$ spectrum for the refereed sample ($m = 68$), obtained using equation 4, revealed a pure phonon characteristic of this mode (no coupling with plasmon exists) and a correlation length ($L_c$) of 0,5nm (hence of the same order of the unit cell). As a matter of fact, for such a small coherence length, the selection rules are not expected to be obeyed indeed. In the case of this long period SL, we than expect no miniband formation.

Referring yet to Fig. 2, as $m$ decreases (short period SL's), significant differences arise in the two polarizations studied, showing that now the selection rules are obeyed. Observing for instance the case for $m = 10$: the

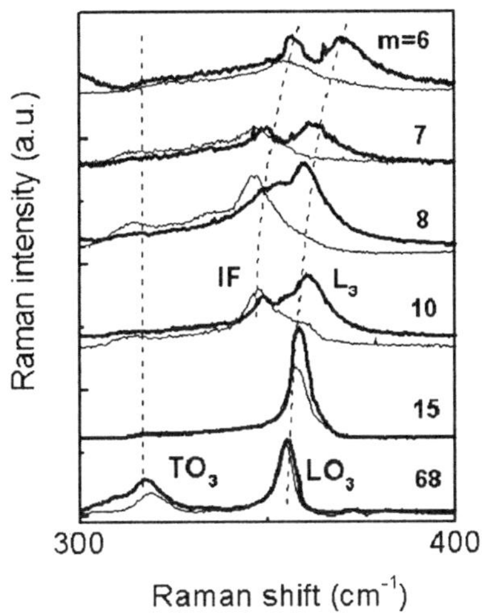


Fig. 2. Details of the Raman spectra of the $(InGaAs)_m(InP)_m$ SL's with different periods in the range of the InP modes. Thin and thick lines correspond to the $z(x', x')\bar{z}$ and $z(x', y')\bar{z}$ polarizations respectively and the dashed lines indicate the positions of the corresponding Raman lines.

$LO_3$ mode is not observed in the $z(x', x')\bar{z}$ spectrum (thin line), in accordance with the Raman selection rules. A line — shifted to high energies — is however observed in the $z(x', y')\bar{z}$ configuration (thick line) that we called $L_3$, and that can be attributed to the coupling between the $LO_3$ and the plasmon excitation mode, present in these samples due to the high density of free carriers. As a matter of fact, in order to achieve the best fitting [4] for this mode ($L_3$) and to obtain its coherence length, it was necessary to consider both the dispersions: that of the pure LO mode and the coupled LO–plasmon one; this further confirms the above attribution made for this line. It is also observed in Fig. 2 an interface (IF) mode for the short period SL's; besides, the $TO_3$ mode is not shifted in the spectra there shown. Moreover, further decreasing the period of the SL (see for instance the case for $m = 6$), the blue shift of $L_3$ increases; the IF mode behaves similarly, thus revealing a longitudinal character. The fittings, obtained using equation 4, have shown that an increase in the coherence length occurs as $m$ decreases; for instance, when $m = 6$, $L_c = 12$nm, much larger than the unit cell size, thus justifying the emergence of the selection rules.

The changes in the spectra (and the plasmon appearance) can be associated to the formation of the minibands due to the SL superperiodicity; in order to confirm this formation, the coupled modes frequencies were calculated. For that, we calculated the effective masses of the electrons propagating in the direction of the confinement $z$ ($m_z$) of the several minibands (for each period of our SL's) using the envelope function approximation [1] with the band parameters taken from Ref. 2. If the minibands are indeed formed, their dispersions (that increase by decreasing the periods of the SL's), should be taken into account; the effective masses should decrease simultaneously, thus resulting in the increase of $\omega_{pz}$. The coupled modes frequencies were then calculated as the zeros of the diagonal component of the dielectric function tensor of a SL in the long wavelength approximation, as foreshown in the introduction section.

Since uniquely the $L_3$ varies substantially with the plasma frequency $\omega_{pz}$, only this mode was used in the frequency calculation. The results are shown in Fig. 3 in terms of the differences between $L_3$ and the pure $LO_3$ phonon (as observed for $m = 68$). The good agreement with the experimental frequency shifts, supports the hypothesis of the minibands formation in the short period SL's.

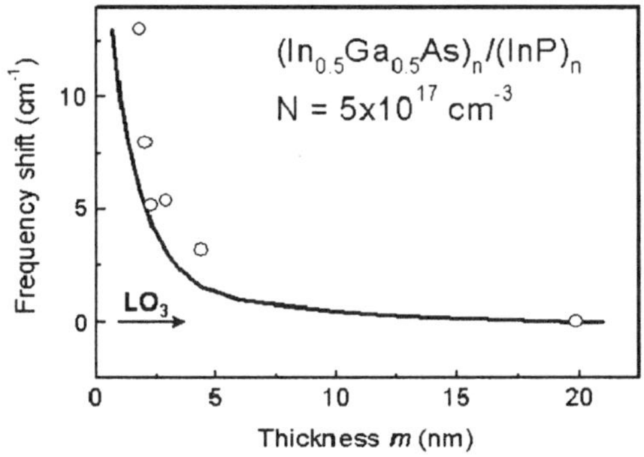


Fig. 3. Calculated (full line) and measured (open circles) values of the differences between the coupled $L_3$ and the pure $LO_3$ modes, as a function of the layer thicknesses.

## References


1. G. Bastard, Phys. Rev. B **24**, 5693 (1981)
2. I. Vurgaftman, J. R. Meyer and L. R. Ram-Mohan, J. Appl.Phys. **89**, 5815 (2001)
3. Yu. Pusep, A. Milekhin, and A. Toropov, Superlatt. Microsctruc. **13**, 115 (1993)
4. H. Richter, Z. P. Wang and L. Ley. Solid State Communications, **39**, 625 (1981)

# ELECTRONIC AND OPTICAL PROPERTIES OF ZnO BETWEEN 3 AND 32 eV

M. Rakel[1]*, C. Cobet[1], N. Esser[1], P. Gori[2],O. Pulci[3], A. Seitsonen[4], A. Cricenti[2], N.H. Nickel[5], W. Richter[3]

[1] *ISAS - Institute for Analytical Sciences, Berlin, Germany,*
[2] *Istituto di Struttura della Materia, CNR, Rome, Italy*
[3] *Dipartimento di Fisica, Roma II (Tor Vergata), Rome, Italy*
[4] *CNRS - IMPC - Université Pierre et Marie Curie, Paris, France*
[5] *Hahn-Meitner-Institut, Berlin, Germany*

The dielectric response functions of bulk ZnO for electric field polarizations parallel and perpendicular to the *c*-axis are obtained in the spectral range from 3 to 32 eV by analysis of ellipsometric data. Anisotropies are observed between $\varepsilon_{\parallel}$ and $\varepsilon_{\perp}$. Electronic transitions involving Zn-3d and O-2s bands are detected. Ab-initio band structure calculations performed at the DFT-LDA level help to interpret the observed transitions. The calculations are also extended to the GW approximation in order to determine the electronic bandgap. Finally, the plasmon frequency is found to be $\hbar\omega_p$=18.95 eV for $E \perp c$ and 18.12 eV for $E \parallel c$, respectively.

*Keywords*: ZnO, ellipsometry, dielectric function, loss function, plasmon, DFT-LDA, GW

## 1. INTRODUCTION

The band gap energy and lattice constants of ZnO are very close to GaN, which was extensively studied in the last decade, but ZnO provides much stronger excitonic features with binding energies around 60 meV. One expects a much higher quantum efficiency for optically pumped room temperature lasing in thin films.[1] The rediscovery is rooted mainly to the recent success in growing high quality single crystals as well as layered films and heterostructures with small mismatch (MgZnCdO-system) and self organized nanorods with a very high purity.

Although many band edge measurements on ZnO exist (first investigations in 1960 by Thomas[2]), ellipsometric spectra with a higher energy

*present address: Paul-Drude-Institut, 10117 Berlin, Germany

range in directions parallel and perpendicular to the optical axis are rare. A major advantage of ellipsometry is the direct and reference free determination of the dielectric function (DF) thus allowing (from its imaginary part) a direct determination of optical transition energies. This avoids the problems occurring in reflectance measurements through application of the Kramers-Kronig analysis and the principal problems occurring in photoluminescence evaluation.[3] Moreover, the DF measured with ellipsometry can be directly converted to the electron loss function $\mathrm{Im}(-1/\varepsilon)$ and is therefore an alternative to EELS (Electron Energy Loss Spectroscopy). Another concern must be the sample quality. Especially in hetero epitaxial layers large strains can exist and might shift the electronic states as well as critical points in the DF.

Theoretical calculations of the electronic band structure indicate that the major structure should occur at energies greater than 10 eV, but most research on optical properties has been limited to energies below 10 eV. The main interband and core level area is therefore not yet investigated by ellipsometry.

This paper reports the ellipsometric measured DF and loss function of single crystal ZnO in the energy range 3-32 eV and at temperatures of 300 K. These results extend the energy range at which the optical properties are known (21 eV from reflectivity measurements[4,5]) and further present an analysis of the given anisotropy along and perpendicular to the *c*-axis of the crystal.

Ab-initio electronic band-structure calculations are also presented. Results obtained within Density Functional Theory (DFT[6]) may help in a first interpretation of experimental data, but they give a large underestimation of the band gap. Calculations performed in the so called GW approximation[7] are therefore also reported for improving the accuracy of the theoretical description.

## 2. EXPERIMENT

The ZnO sample is a commercially available single crystal cleaved at the *a*-plane to achieve a surface normal along the $[11\bar{2}0]$ direction. The orientation of the *c*-axis in the surface plane was determined by reflectance anisotropy spectroscopy (RAS). The RAS determination of anisotropy is also very helpful to verify the ellipsometric data (for detailed description see Ref.[8]).

All measurements were performed under ultrahigh-vacuum conditions with a rotating analyzer ellipsometer by using a normal incidence monochromator (2.5-10 eV) and a toroidal grating monochromator (8-

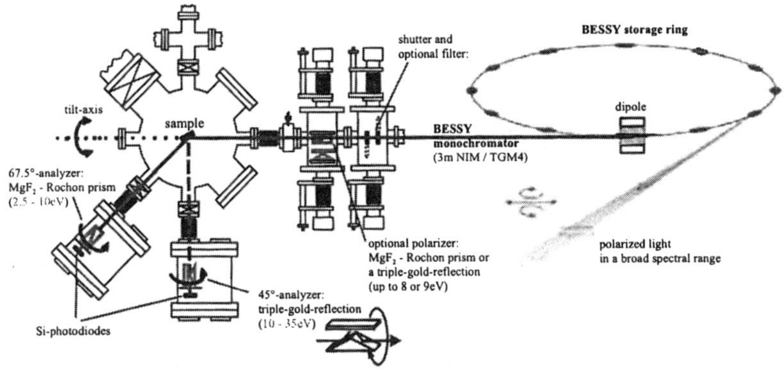


Fig. 1. Sketch of the VUV-XUV-Ellipsometer, which was designed to operate with synchrotron radiation. It consists of the main chamber with a sample manipulator and two chambers with rotating analyzers. Additionally, there is the polarizing chamber with a pinhole alignment system and a filter chamber with some optional filters and a shutter. All components including motors and sample are in UHV.

35 eV) beamline at BESSY II. A sketch can be seen in figure 1. The polarization vector of the beam was tilted to 20° with respect to the plane of incidence during the ellipsometric measurements. A $MgF_2$-Rochon prism was used as polarizer and analyzer for energies below 10 eV. Since no birefringent transparent materials transmit light above 10 eV, the incoming monochromized synchrotron radiation was then linearly polarized by a triple gold reflection polarizer. The linearly polarized part of this polarizer is calculated to 98.5%. Together with the native linear polarization of the synchrotron light, thus, we can assume a degree of polarization considerably higher than this value. The synchrotron light provided by a bending magnet contains radiation from the far infrared up to hard X-rays. Thus, the diffracted light from the monochromator grating appears in several orders. Unfortunately, it was not possible to suppress contributions of second order light between 10 and 14 eV due to the lack of suitable filters. This is a well known problem. In the future the use of very thin rare earth foils is planned.

The high energy limit depends on the reflectivity of the samples. Above the plasma edge of the material the reflectivity decreases rapidly and is nearly zero at 32 eV at 45° angle of incidence.

The use of synchrotron radiation supplies an intense polarized continuum spectrum of light throughout the vacuum ultraviolet region.

## 3. COMPUTATIONAL METHOD

The ab-initio calculations performed in this work are based on Density Functional Theory in the Local Density Approximation (LDA). Since it is well known that band gaps are underestimated within DFT, quasiparticle corrections to DFT energies are calculated in the GW approximation.

The lattice parameters employed in the calculation are $a$=3.191 Å; $c$=5.153 Å; $u$=0.379. The electron-ion interaction is described through norm-conserving pseudopotentials (PP). For Zn, three different PP have been employed: one with only 4$s$ electrons and with nonlinear core corrections ($Zn^{2+}$ PP), another one with also 3$d$ electrons ($Zn^{12+}$ PP), and a third in which all the electrons in the third shell are treated as valence states ($Zn^{20+}$ PP). The wavefunctions are expanded in plane waves with a kinetic cutoff energy equal to 60, 70 and 200 Ry for the three types of used pseudopotentials. GW calculations have been performed with 324$k$ points in the BZ and using 180 empty bands.

## 4. RESULTS

**A. Electronic band structure and DOS:** The band structure and density of states obtained from DFT-LDA (using the $Zn^{12+}$ PP) is displayed in figure 2. Although the 3$d$-level of Zn strongly overlaps with the upper valence bands and therefore takes part in the chemical bonding too, the localization (weak dispersion) is still noticeable. Strong photon absorption should occur at energies about 12-15 eV by exciting these $d$-electrons to the lowest $p$-like conduction states .

A comparison between the band gaps calculated within DFT and GW, and with the three types of Zn PPs, can be made by analyzing the values listed in table 1. Even if it appears that the use of a $Zn^{2+}$ PP (that is the pseudopotential without $d$ states) provides the gap which is closer to the experimental value of 3.44 eV, this agreement has to be considered fortuitous since it has been shown[9] that in this approach lattice constants are underestimated and energy eigenvalues do not fit well with experiments. Given the energy position of Zn 3$d$ states (between -6.5 and -4 eV, as shown in figure 2), these have clearly to be taken into account for a realistic description of the system. However, using a $Zn^{12+}$ PP, the gap opens only marginally with GW and is far from the experimental value. The inclusion of the whole third shell in the Zn PP improves somewhat the situation, but still the gap at $\Gamma$ remains underestimated by about 0.9 eV.

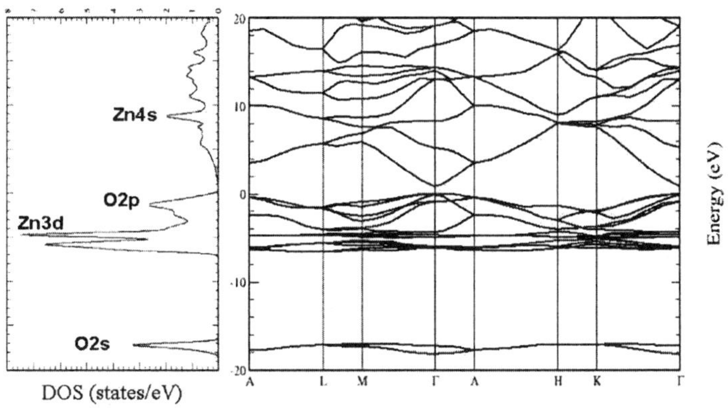


Fig. 2. DOS and electronic band structure of ZnO as obtained from DFT-LDA calculations. The labels describe which kind of orbital of the atomic wavefunctions mainly contribute to the total DOS.

Table 1. LDA and GW gaps at high-symmetry points. Energies are in eV.

| | $Zn^{2+}$ PP | | $Zn^{12+}$ PP | | $Zn^{20+}$ PP | |
|---|---|---|---|---|---|---|
| k-point | LDA | GW | LDA | GW | LDA | GW |
| $\Gamma$ | 1.94 | 3.60 | 0.81 | 0.83 | 0.63 | 2.52 |
| $A$ | 4.95 | 6.91 | 3.93 | 4.34 | 3.75 | 5.90 |
| $H$ | 9.85 | 11.91 | 9.46 | 10.61 | 9.23 | 11.78 |
| $K$ | 10.14 | 12.08 | 9.55 | 10.48 | 9.31 | 11.54 |

**B. Interband and core level transitions:** In the energy region from 3 to 7 eV no structure due to critical points in the interband transitions is observed (figure 3). This is different from all other II-VI materials with wurtzite structure, probably due to the energetic position of the Zn $3d$ level. The interband spectra of wurtzite crystals are indeed predicted to exhibit anisotropic behavior, but ZnO is strongly different from e.g. GaN or InN, where this anisotropy is governed by symmetry selection rules and can be directly explained within the zincblende - wurtzite analogy.[10,11] This could be related to a stronger deviation of the internal parameter $u$ shown by ZnO ($u_{\mathrm{GaN}} = 0.377$, $u_{\mathrm{InN}} = 0.379$, $u_{\mathrm{ZnO}} = 0.382$) from the value $u_{\mathrm{ideal}} = 0.375$ of an ideal wurtzite crystal.

We observe that the DF for ZnO (figure 3) does not show remarkable anisotropy. The most sizeable anisotropy appears below 7 eV. The peak around 3.4 eV is due to excitonic effects at the fundamental gap. Around

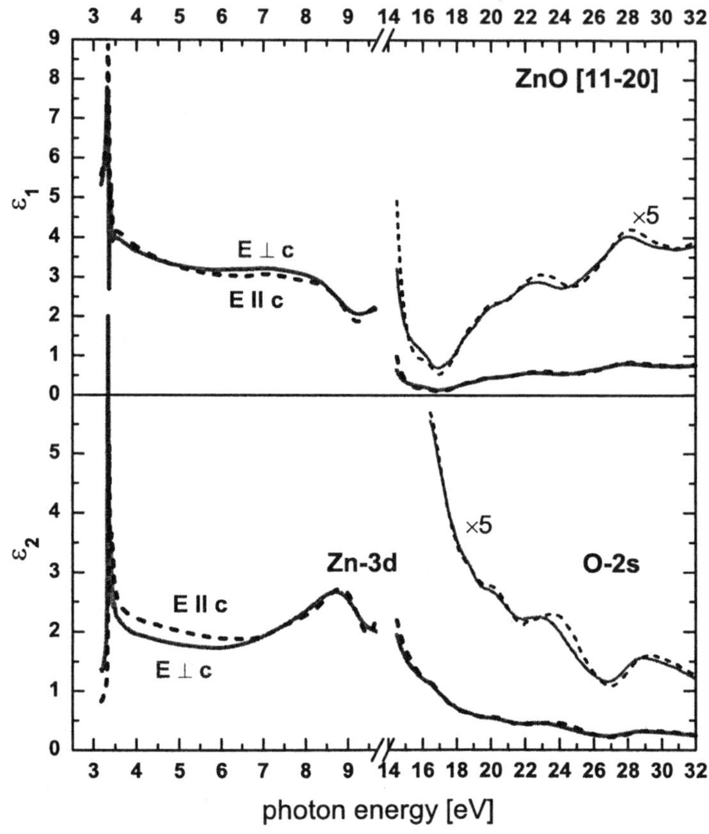


Fig. 3. Real and imaginary part of the dielectric function parallel and perpendicular to the optical $c$-axis. The labels suggest an assignment of electronic transitions occurring around high symmetry points in the Brillouin zone.

9 eV we observe a strong peak which involves Zn $3d$ states. At higher energies the absorption decreases, but shows some residual structures and anisotropies in the energy range of O $2s$ level.

**C. Loss function:** The energy loss function is represented by the imaginary part of the reciprocal dielectric function $\mathrm{Im}(-1/\varepsilon)$ and can be directly observed by ellipsometry, too. While optical techniques can only access the transversal DF $\varepsilon_T$, the longitudinal excitations in $\varepsilon_L$ are, in general, not reached. But, in the long wavelength limit, these two functions are equivalent. Since the wavelengths of photons with energies 14-32 eV used in this work are much larger than the distances of atoms in the lattice, the measurements are assumed to approximate this limit.

The global maximum in the energy loss spectrum in the range 14-32 eV (figure 4) corresponds to values in $\varepsilon_1$ and $\varepsilon_2$ which are near zero and can be interpreted as a single true plasmon excitation. This collective longitudinal oscillation with a frequency $\hbar\omega_p$ of valence electrons appears at the border of valence electron transitions, where $n_{\text{eff}}$ (the number of available electrons for optical excitation) is nearly exhausted. Going higher in energy only the two O2$s$ electrons are expected to additionally contribute to the collective excitation. We deduce a plasmon frequency $\hbar\omega_p$ of 18.12 eV for directions

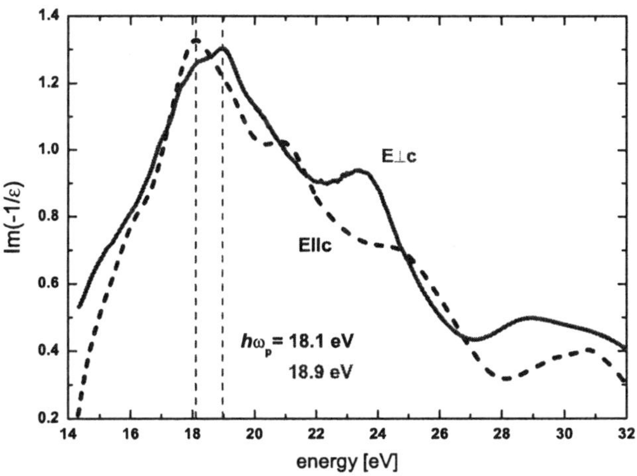


Fig. 4. Energy loss function of ZnO resolved parallel and perpendicular to the $c$-axis. The plasmon energy is given by the global maximum

perpendicular and 18.95 eV for directions parallel to the $c$-axis. These values agree with electron energy loss (EELS) data: 18.8 eV for fields $\parallel$ to $c$ from[5] and 18.5 eV averaged over both,[4] but are much more accurate.

The loss function calculated at the DFT level ($Zn^{12+}$ PP) is shown in figure 5. The global maxima are 15.1 eV for $E \parallel c$ and 15.5 eV for $E \perp c$. The underestimation of the measured plasma frequency is somewhat larger than that caused by the DFT gap underestimation and is probably related to the complex dependence of the loss function on both $\varepsilon_1(\omega)$ and $\varepsilon_2(\omega)$. Apart from these differences in peak positions, the overall shape of the calculated loss function agrees reasonably with the measured one and a clear anisotropy at energies between 18 and 22 eV is detected.

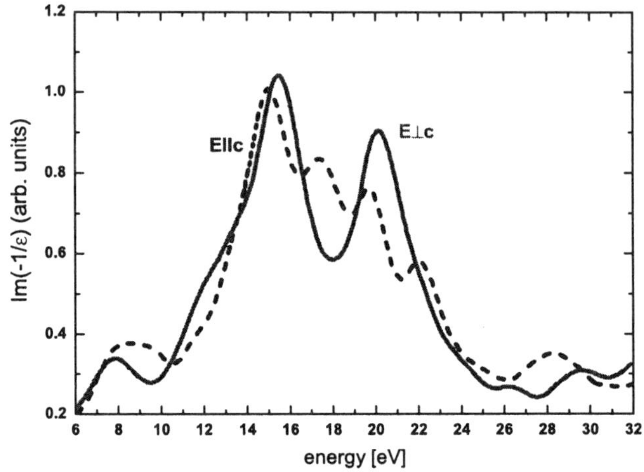


Fig. 5. Calculated loss function of ZnO parallel and perpendicular to the $c$-axis.

## 5. CONCLUSIONS

The complex dielectric tensor of ZnO was measured by ellipsometry in the whole range of interband transitions. Anisotropy is observed especially near the band edge. A strong excitonic peak is found around 3.4 eV. Residual anisotropy is present also in the energy range of O2$s$ level. Also the energy loss function, both measured and calculated, presents a clear anisotropy above the plasma frequency.

## 6. ACKNOWLEDGEMENTS

We are grateful to BESSY and co-workers. We thank the BMBF for funding under 05 KS4KTB/3. CPU time has been granted by CINECA. Financial support by MIUR through PRIN 2004 (2004028932) is also acknowledged.

### References

1. D.M. Bagnall, Y.F. Chen, Z. Zhu, S. Koyama, and T. Goto, Appl.Phys. Lett. **70**, 2230 (1997)
2. D.G. Thomas, J. Phys. Chem. Solids **15**, 86, (1960)
3. W. Shan, W. Walukiewitz, J.W. Ager, and K.M. Yu, Appl. Phys. Lett. **86**, 191911 (2005)

4. R. Klucker, H. Nelkowski, Y.S. Park, M. Skibowski, and T.S. Wagner, phys. stat. sol. (b) **45**, 265 (1971)
5. R.L. Hengehold and R.J. Almassy, Phys. Rev. **B 1**, 4784 (1970)
6. P. Hohenberg and W. Kohn, Phys. Rev. **136**, B864 (1964).
7. F. Aryasetiawan and O. Gunnarsson, Rep. Prog. Phys. **61**, 237 (1998).
8. U. Rossow, R. Goldhahn, D. Fuhrmann, and A. Hangleiter, phys. stat. sol (b), 242 No 13 (2005)
9. P.Schroer, P. Kruger, and J. Pollmann, Phys. Rev. B **47**, 6971 (1993).
10. M. Cardona, and G. Harbeke, Phys. Rev. **137**, A1467 (1965).
11. Christoph Cobet, *Linear Optical Properties of III-Nitride Semiconductors between 3 and 30 eV*, Dissertation at TU-Berlin (2005) available at: http://opus.kobv.de/tuberlin/volltexte/2005/1176/

# ORDER AND CLUSTERS IN MODEL MEMBRANES: DETECTION AND CHARACTERIZATION BY INFRARED SCANNING NEAR-FIELD MICROSCOPY


**J. Generosi, G. Margaritondo**
Institut de physique appliquée, Ecole Polytechnique Fédérale, CH-1015 Lausanne, Switzerland
**J. S. Sanghera, I. D. Aggarwal**
Optical Sciences Division, U.S. Naval Research Laboratory, 4555 Overlook Ave SE, Washington, DC 20375, USA
**N. H. Tolk**
Department of Physics and Astronomy, Vanderbilt University, Nashville, TN, 31235, USA
**D. W. Piston**
Department of Molecular Physiology and Biophysics, Vanderbilt University, Nashville, TN 37232, USA
**A. Congiu Castellano**
Dipartimento di Fisica, Università La Sapienza, Piazzale A. Moro 2, 00185 Roma, Italy
**A. Cricenti**
Istituto di Struttura della Materia, CNR, Via Fosso del Cavaliere 100, 00133 Roma, Italy and Department of Physics and Astronomy, Vanderbilt University, Nashville, TN, 31235 USA



**Lipid bilayers are critical components of the cell and can be considered good model systems for cell membranes. Order, stability and clustering of these components are crucial parameters in many biotechnological and medical applications, such as biosensors and gene therapy. Glass-supported 1,2-dioleoyl-sn-glycero-3-phosphocholine (DOPC) neutral lipid membranes were studied by infrared scanning near-field optical microscopy (IR-SNOM) at several wavelengths, obtaining a mapping of the chemical contents of the sample. Optical micrographs indicate the formation of locally ordered multibilayers and topographical images reveal the presence of micron-sized islands at the surface.**


## INTRODUCTION

Supported lipid bilayers can be considered a versatile model system for investigating a wide variety of phenomena. Their possible application in medical and biotechnological fields [1, 2] have led many researchers to investigate the fine structure of the lipid components by infrared and Raman spectroscopy [3], the structural properties of bilayers through small angle X-ray scattering [4, 5], X-ray and neutron reflectivity [6], membrane chemical structure or physical properties with fluorescence spectroscopy [7-10], as well as cryoelectron and atomic force microscopies [11-15]. However, a direct visualization of ordering and clustering of lipids is still lacking.

Due to its surface sensitivity and high spatial resolution, scanning near-field optical microscopy (SNOM) has a significant potential to study the lateral organization of membrane domains and clusters. SNOM can be combined to several spectroscopic techniques; among these, fluorescence and infrared absorption spectroscopy are the most popular ones, especially for biological applications [16, 3]. Compared to other techniques, infrared near-field microscopy in the spectroscopic mode has the advantage to be sensitive to specific chemical bonds; spectroscopic SNOM in the infrared spectral range (IR-SNOM) can reveal the chemical content of the sample with a lateral resolution around 100 nm [17, 18], well beyond the classical diffraction limit [19, 20].

We present a topographical and spectroscopic IR-SNOM study of neutral lipid multibilayers deposited on glass: several wavelengths, corresponding to the vibrational modes of the interested chemical bonds, have been used to obtain maps of the chemical contents of the sample.

## MATERIALS AND METHODS

The lipid 1,2-dioleoyl-sn-glycero-3-phosphocholine (DOPC), purchased from Avanti Polar Lipids, Alabaster, AL, USA, was dissolved in chloroform; the resulting organic phase was evaporated at 50°C in a rotary evaporator. The thin dried film was placed under vacuum to ensure that all traces of solvent had been removed. After hydration by deionised water the solution was sonicated for 20 min. The solution was then deposited on glass surfaces. In fact, immediate condensation into bilayers had been observed for DOPC vescicles deposed on glass [21]. Structural order and lamellar phase of the samples was confirmed by x-ray and neutron reflectivity measurements [6].

The unfocused Vanderbilt Free Electron Laser beam is produced by a 45 MeV radiofrequency accelerator. The source is tunable between 2 and 10 μm of wavelength. The IR source illuminates the sample and the near-field probe collects the reflected light. For each desired wavelength, the specially designed SNOM module (explained elsewhere by Cricenti et al. [22]) collects an optical image simultaneously with the topographical shear-force micrograph of the scanned area.

Infrared SNOM probes were obtained from single-mode, one-meter-long, arsenic selenide fibers having a 50-μm-core diameter. One end of the fiber was interfaced to an InSb detector while the remaining end of the fiber was previously chemically etched [23] and coated with gold with an approximate thickness of 100-125 nm.

# RESULTS AND DISCUSSION

Lipid membranes were studied by IR-SNOM at several wavelengths, corresponding to several absorption (5.75, 6.8 and 8.05 μm) and to the background (6.1 μm). Topographical micrographs reveal the presence of islands at the surface and the optical images indicate the formation of locally ordered multiple bilayers.

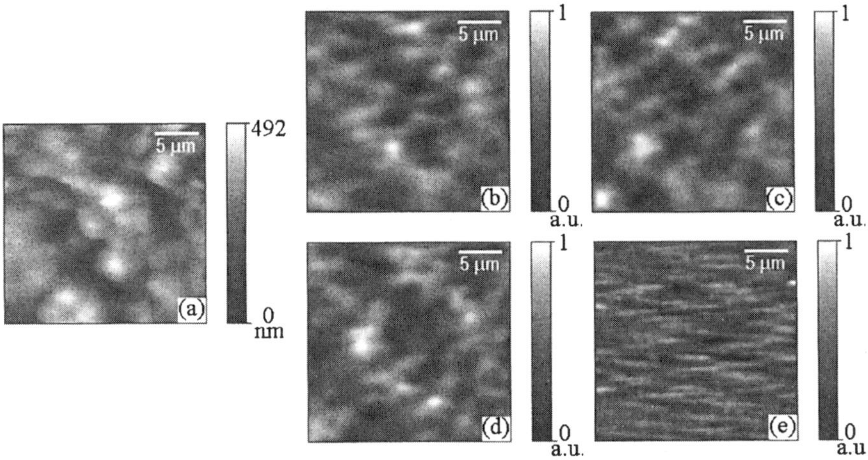


Figure 1: (a) 20 μm x 20 μm topographical (shear-force) micrograph of solid-supported neutral lipid. (b),(c), (d) and (e) Chemical mapping at different wavelengths (8.05, 6.8, 5.75 and 6.1 μm).

Figure 1(a) shows a 20x20 $\mu m^2$ area where the surface is not homogenous: clusters of several hundred of nanometers are clearly visible. This is consistent with AFM studies [24] that showed a similar topography, demonstrating the existence of a lamellar structure in the case of a nonhomogeneous dried lipid films. The tendency of the lipid vescicles to fuse together depends on many parameters, discussed in several works [25]. Nonuniform structures [26] can be obtained by simple drying of a droplet of the lipid suspension placed on glass: individual lipid vesicles can partly fuse together and complex with the bilayers below [27].

In order to map the chemical distribution of the lipids, the FEL beam was tuned at four specific wavelengths: λ=5.75, 6.1, 6.8 and 8.05 μm. Figure 1(b) is taken at 8.05 μm corresponding to the absorption of the $PO2^-$ vibrational group, which is present only in the "head" of the neutral lipid, while Figures 1(c) and 1(d) show SNOM images taken at 6.8 and 5.75 μm that correspond to the absorption of CH2 and CO groups, in the "tails" of the DOPC. The spatial distribution of these is shown by the darkest spots in the optical micrographs (b), (c), (d) and (e) of Figure 1. The intensity of each pixel was normalized by the corresponding measured FEL intensity; all images were normalized to have the same dynamic range.

The optical micrographs (Fig. 1 (b), (c) and (d)) show similar absorption patterns: tail and head of the lipid are aligned in the z direction. We can assume that islands of lamellar-structured bilayers are formed, even though these clusters are not fused together to form a flat surface.

The image of Figure 1(d) was taken at 6.1 μm where there is no absorption from DOPC. The micrograph is structureless and shows only background noise; this confirms that optical images are not affected by topographic artefacts.

## CONCLUSIONS

We performed a topographical and a chemical characterization of the solid-supported neutral lipid DOPC by IR-SNOM. Near-field micrographs corresponding to different vibrational bands did reveal locally ordered multiple bilayers and micron-size clusters in the glass-supported lipid membrane. DOPC multibilayers are highly ordered along the z-axis whereas in the x-y plane they consist of ordered island separated by amorphous regions. Our conclusions are supported by x-ray data [28], previous neutron reflectivity [6] and atomic force microscopy studies [24]. Such properties are crucial for possible applications in the biotechnology field, for example with biosensors, and in medicine, with gene therapy.

## ACKNOWLEDGMENTS

We gratefully acknowledge the support of the Vanderbilt University FEL Center, operating under the Department of Defense MFEL Program. The experiments were supported by the Italian National Research Council and by the Fonds National Suisse de la Recherche Scientifique.

## References

[1] E. Sackmann, "Supported membranes: scientific and practical applications", *Science*, vol. 271, pp. 43-48, 1996.

[2] C. R. Safinya, "Structures of lipid–DNA complexes: supramolecular assembly and gene delivery", *Curr. Opin. Struct. Biol.*, vol. 11, pp. 440-448, 2001.

[3] C. S. Braun, G. S. Jas, S. Choosakoonkriang, G. S. Koe, J. G. Smith and C. R. Middaugh, "The structure of DNA within cationic lipid/DNA complexes", *Biophys. J.*, vol. 84, pp. 1114-1123, 2003.
[4] I. Koltover, T. Salditt, J. O. Rädler, and C. R. Safinya, "An inverted hexagonal phase of cationic liposome-DNA complexes related to DNA release and delivery", *Science*, vol. 281, pp. 78-81, 1998.
[5] J. O. Radler, I. Koltover, T. Salditt, and C. R. Safinya, "Structure of DNA-cationic liposome complexes: DNA intercalation in multilamellar membranes in distinct interhelical packing regimes", *Science*, vol. 275, pp. 810-814, 1997.
[6] J. Generosi, C. Castellano, D. Pozzi, R. Felici, G. Fragneto, F. Natali and A. Congiu, "X-ray and neutron reflectivity study of solid-supported lipid membranes prepared by spin coating", *J. Appl. Phys.*, vol. 96, pp. 6839-6844, 2004.
[7] H. Gershon, R. Ghirlando, S. B. Guttman and A. Minsky, "Mode of formation and structural features of DNA-cationic liposome complexes used for transfection", *Biochem.*, vol. 32, pp.7143-7151, 1993.
[8] N. J. Zuidam and Y. Barenholz, "Electrostatic and structural properties of complexes involving plasmid DNA and cationic lipids commonly used for gene delivery", *Biochim. Biophys. Acta*, vol. 1368, pp. 115-128, 1998.
[9] D. Hirsch-Lerner and Y. Barenholz, "Probing DNA-cationic lipid interactions with the fluorophore trimethylammonium diphenyl-hexatriene (TMADPH)", *Biochim. Biophys. Acta*, vol. 1370, pp. 17-30, 1998.
[10] S. Bhattacharya and S. S. Mandal, "Evidence of interlipidic ion-pairing in anion-induced DNA release from cationic amphiphile-DNA complexes. Mechanistic implications in transfection", *Biochem.*, vol. 37, pp. 7764-7777, 1998.
[11] Y. Xu, S. W. Hui, P. Frederik and F. C. Jr Szoka, "Physicochemical characterization and purification of cationic lipoplexes", *Biophys. J.*, vol. 77, pp. 341-353, 1999.
[12] C. Kawaura, A. Noguchi, T. Furuno and M. Nakanishi, "Atomic force microscopy for studying gene transfection mediated by cationic liposomes with a cationic cholesterol derivative", *FEBS Lett.*, vol. 421, pp. 69-72, 1998.
[13] V. Oberle, U. Bakowsky, I. S. Zuhorn and D. Hoekstra, 2Lipoplex formation under equilibrium conditions reveals a three-step mechanism", *Biophys. J.*, vol. 79, pp. 1447-1454, 2000.
[14] F. Sakurai, T. Nishioka, H. Saito, T. Baba, A. Okuda, O. Matsumoto, T. Taga, F. Yamashita, Y. Takakura and M. Hashida, "Interaction between DNA-cationic liposome complexes and erythrocytes is an important factor in systemic gene transfer via the intravenous route in mice: the role of the neutral helper lipid", *Gene Ther.*, vol. 8, pp. 677-686, 2001.
[15] S. Huebner, B. J. Battersby, R. Grimm and G. Cevc, "Lipid-DNA complex formation: reorganization and rupture of lipid vesicles in the presence of DNA as observed by cryoelectron microscopy". *Biophys. J.*, vol.76, pp. 3158-3166, 1999.

[16] D. Vobornik et al., "Infrared near-field microscopy with the Vanderbilt Free Electron Laser: overview and perspectives", *Infrared Physics & Technology*, vol. 45, pp. 409-416, 2004.
[17] A. Cricenti, R. Generosi, P. Perfetti, J. M. Gilligan, N. H. Tolk, C. Coluzza, and G. Margaritondo, "Free-electron-laser near-field nanospectroscopy", *APL*, vol. 73, pp. 151-153, 1998.
[18] A. Cricenti et al., "Chemically resolved imaging of biological cells and thin films by infrared scanning near-field optical microscopy", *Biophys. J.*, vol. 85, pp. 2705-2710, 2003.
[19] D. W. Pohl, W. Denk, M. Lanz, "Optical stethoscopy: Image recording with resolution λ/20", *Appl. Phys. Lett.*, vol. 44, pp. 651-653, 1984.
[20] R.C. Dunn, "Near-field scanning optical microscopy", *Chem. Rev.*, vol. 99, pp. 2891-2928, 1999.
[21] M. Benes, D. Billy, A. Benda, H. Speijer, M. Hof and W. Th. Hermens, "Surface-dependent transitions during self-assembly of phospholipid membranes on mica, silica, and glass", *Langmuir*, vol. 20, pp. 10129-10137, 2004.
[22] A. Cricenti, R. Generosi, C. Barchesi, M. Luce, M. Rinaldi, "A multipurpose scanning near-field optical microscope: reflectivity and photocurrent on semiconductor and biological samples", *Rev. Sci. Instrum.*, vol. 69, pp. 3240-3244, 1998.
[23] D. B. Talley, L. B. Shaw, J. S. Sanghera, I. D. Aggarwal, A. Cricenti, R. Generosi, M. Luce, G. Margaritondo, J. M. Gilligan and N. H. Tolk, "Scanning near field infrared microscopy using chalcogenide fiber tips", *Mater. Lett.*, vol. 42, pp. 339-344, 2000.
[24] G. Pompeo, M. Girasole, C. Cricenti, F. Cattaruzza, A. Flamini, T. Prosperi, J. Generosi, A. Congiu Castellano, "AFM characterization of solid-supported lipid multilayers prepared by spin-coating", *Biochim. Biophys. Acta*, vol. 1712, pp. 29-36, 2005.
[25] I. Reviakine and A. Brisson, "Formation of supported phospholipid bilayers from unilamellar vesicles investigated by atomic force microscopy", *Langmuir*, vol. 16, pp. 1806-1815, 2000 and reference therin.
[26] A. Cohen Simonsen and L. A. Bagatolli, "Structure of spin-coated lipid films and domain formation in supported membranes formed by hydration", *Langmuir*, vol. 20, pp. 9720-9728, 2004.
[27] S. Kumar and J. H. Hoh, "Direct visualization of vesicle-bilayer complexes by atomic force microscopy", *Langmuir*, vol. 16, pp. 9936-9940, 2000.
[28] J. Generosi, G. Pompeo, R. Felici, C. Castellano, F. Domenici and A. Castellano, *Unpublished data*.

# CHEMICAL AND MAGNETIC PROPERTIES OF NiO THIN FILMS EPITAXIALLY GROWN ON Fe(001)

ALBERTO BRAMBILLA

*Dipartimento di Fisica del Politecnico di Milano,*
*Piazza Leonardo Da Vinci 32, 20133 Milano, Italy*
*E-mail: alberto.brambilla@polimi.it*

Very high quality NiO films have been grown on Fe(001) by means of Molecular Beam Epitaxy. The chemical and magnetic properties of the NiO/Fe(001) interface have been evaluated by means of X-ray Absorption Spectroscopy and X-ray Magnetic Circular Dichroism. Furthermore, combined use of X-ray Magnetic Linear Dichroism and PhotoElectron Emission Microscopy allowed to observe an in-plane uniaxial magnetic anisotropy in very thin NiO films. For NiO films thinner than about 9 atomic layers the NiO magnetic moments align in-plane perpendicular to the Fe substrate magnetization. Above such critical thickness the coupling turns out to be collinear. The effects of thermal treatments, fundamental to produce exchange-biased structures, have also been considered.

## 1. Introduction

Oxide thin films and metal-oxide interfaces play a fundamental role in many chemical-physics processes and applications in material science[1]. In particular, if an antiferromagnetic (AF) oxide, is coupled to a ferromagnetic (F) metal, a rich phenomenology at the AF/F interface occurs, related to the Exchange Bias (EB) mechanism. This consists in the breaking of time reversal symmetry and the onset of unidirectional magnetic anisotropy in a field-cooled AF/F interface, with several possible technological applications[2]. In this context, a comprehensive knowledge of the magnetic and chemical properties of the AF/F interface is crucial to determine the type of coupling between the F and AF anisotropy axes and the magnitude of the EB effect.

Among AF/F systems, the NiO/Fe interface has attracted considerable attention[3,4,5], because of the high NiO Néel temperature ($T_N$ = 520 K), and the low lattice mismatch between Fe and NiO[6].

In this work, synchrotron-light based techniques will be exploited to study the chemical and magnetic properties of the NiO/Fe interface. X-ray Absorption Spectroscopy (XAS) has been used to characterize the chemical composition of the buried interface, where oxidation/reduction processes will be shown to occur. X-ray Magnetic Circular Dichroism (XMCD) has given information on the magnetic properties of the so formed interface, while X-ray Magnetic Linear Dichroism (XMLD) has been applied to investigate the magnetic anisotropy in the NiO layer. The latter technique can indeed yield information even in AF systems, where no net magnetization is present, at variance from XMCD, which is sensible to the total magnetization. Furthermore, the secondary electrons produced in the absorption process can be imaged by the PhotoElectron Emission Microscopy (PEEM) technique[7], which allows a spatial resolution of about 100 nm.

## 2. Experiment

The samples were prepared in ultrahigh vacuum (UHV), growing a thick (100 nm thickness range) iron film by means of molecular beam epitaxy on UHV-cleaned MgO(001) single crystal substrates. The Fe film was then exposed to oxygen, in order to obtain the very stable and well characterized Fe(001)-p(1×1)O surface[8].

High quality epitaxial NiO films (1 nm thickness range) with no defect states in the gap can be prepared by reactive deposition, evaporating Ni from a high purity rod in a $8 \times 10^{-7}$ mbar $O_2$ atmosphere on the Fe(001)-p(1×1)O substrate kept at room temperature[6].

It is worth noting that the thickness of the NiO and Fe layers is such that the former is antiferromagnetic and the latter ferromagnetic at room temperature. In each case, the Fe substrate was remanently magnetized along one of its in-plane easy axis, before growing NiO.

The details of the PEEM measurements are described elsewhere[9].

## 3. Chemistry and Magnetism at the NiO/Fe Interface

The XAS and XMCD signals obtained from Fe(001)-p(1×1)O and 20 ML (monolayers, 1 ML NiO = 2.09 Å) NiO/Fe(001) are displayed in Fig. 1a). The top spectrum shows clear fingerprints of oxidation of the Fe substrate, visible as an atomic multiplet fine structure on both the XAS and XMCD signals and indicating a re-localization of the electronic states. Because of the large x-rays probing depth, the substrate contribution to the 20 ML NiO/Fe(001) XAS and XMCD spectra at the Fe $L_{2,3}$ edge is considerable. In order to obtain the spectra of the Fe oxide at the interface, in Fig. 1a) the contribution from the substrate has been subtracted after a rescaling. In the lower part of Fig. 1a), the difference spectrum is compared with the XAS

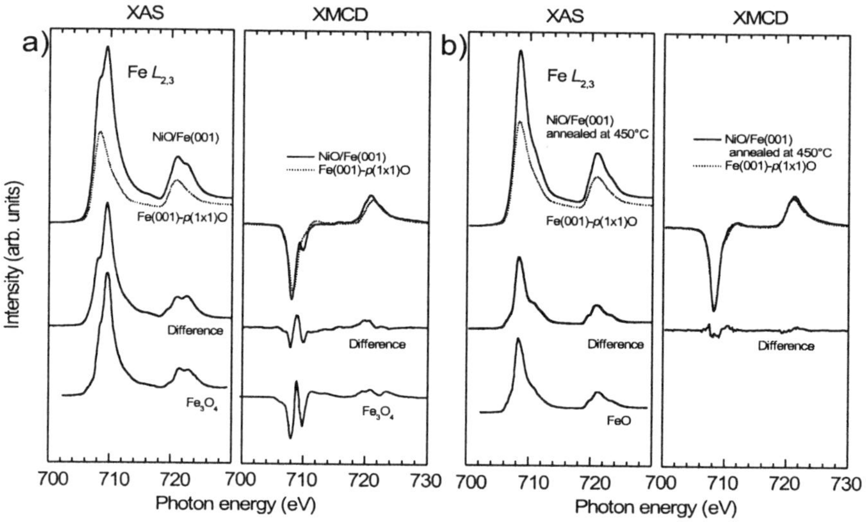


Figure 1. XAS and XMCD spectra for 20 ML NiO/Fe(001) for: a) as grown interface; b) after 450°C annealing.

and XMCD spectra obtained from bulk $Fe_3O_4$, which is ferrimagnetic. The agreement for both the XAS and XMCD line shapes is apparently so good over the entire energy range that the iron oxidation state at the interface can be obviously identified as the ferrimagnetic $Fe_3O_4$ oxide. Reactive deposition of Ni on Fe(001)-p(1×1)O leads to very similar results, in terms of iron oxidation state, as room temperature Fe exposure to $O_2$ which leads to the formation of $Fe_3O_4$ as well[10,11]. A general result in fact is that $Fe_3O_4$ is formed when oxygen is available in large quantities. However, an important difference is that the $Fe_3O_4$ formed after exposure of iron to $O_2$ is coupled antiferromagnetically to the underlying metal[11]. On the contrary, the data collected reveal a ferromagnetic coupling between the substrate and the oxide. This implies that the energetics of a $Fe_3O_4$ film magnetically coupled with an iron substrate by the exchange interaction is dramatically altered by the presence of the AF NiO overlayer. Note also that, although the shape of the difference and $Fe_3O_4$ XMCD spectra are in good agreement, the magnitude of the dichroism in the difference is about a factor two less than the one measured on bulk $Fe_3O_4$, indicating a loss of magnetization in the interface oxide. Furthermore, this analysis clearly shows that the iron oxide that forms at the Fe/NiO interface depends on the growth sequence.

In fact, the interface oxide that results when Fe is deposited on NiO has been identified with FeO[10], which is antiferromagnetic. From the comparison in Fig. 1, an estimate of the interface oxide thickness t can be deduced on the basis of a simple attenuation model[12], obtaining t = $(19 \pm 3)$ Å.

The effects of annealing on the buried interface obtained after reactive NiO deposition have also been investigated. The knowledge of the interface behavior upon annealing is particularly important in the physics of exchange biased systems. Similarly to Fig. 1a), in Fig. 1b) the substrate contribution is subtracted from the NiO/Fe(001) sample annealed at 450°C. The XAS difference spectrum is very similar to the one measured on bulk FeO[10], while the XMCD difference spectrum vanishes as expected for an antiferromagnetic compound such as FeO. In this case, the oxide thickness has been extimated to be t = $(8 \pm 3)$ Å.

Duò *et al.*[6] demonstrated that, in these films, annealing has a dramatic effect on the NiO overlayer as well. A Ni-Fe substitution is indeed found to occur, leading to the formation of iron oxide at the surface[6].

## 4. Magnetic anisotropy in the NiO layer

The absorption spectra for XMLD were obtained measuring the drain current from the sample in four different geometries. The angle between the surface normal and the polarization of the incoming light could have the values $\theta = 90°$ (normal incidence) or $\theta = 30°$ (grazing), while the Fe substrate magnetization **M** was kept parallel ($\mathbf{M} \parallel \mathbf{E}$) or perpendicular ($\mathbf{M} \perp \mathbf{E}$) to the photon electric field **E**. Figure 2 (left side) shows the Ni $L_2$ spectra collected in the four geometries for a 7 ML-thick NiO/Fe(001) film. These spectra are normalized to unity at the low photon energy peak, after subtraction of a background. The spectra reported clearly show a strong dependence with the angle $\theta$ as well as with the relative orientation of **M** and **E**. The differences observed between the spectra measured at normal incidence rotating the azimuth can be attributed to the presence of a long-range antiferromagnetic order with uniaxial anisotropy in the plane of the NiO film, whose direction is perpendicular to that of the magnetization of the iron substrate. This result is surprisingly different from previous experimental results on similar interfaces[4,5], and a possible explanation based on a phenomenological model has been given[13].

The data collected at grazing incidence could be influenced by the uniaxial crystal field, at variance with the normal incidence case where the NiO(001) surface normal is a fourfold symmetry axis for the crystal

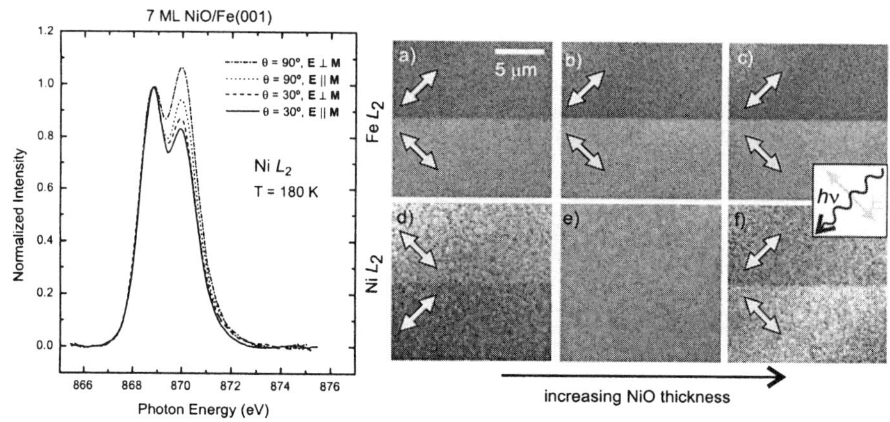


Figure 2. Left: normalized Ni $L_2$ absorption spectra obtained at 180 K on a 7 ML-thick NiO sample grown on Fe(001). Right: PEEM images of XMLD asymmetry obtained at the Fe (top) and Ni (bottom) $L_2$ edge on a NiO/Fe(001) wedged thin film. The crystallographic directions refer to the Fe substrate. The NiO thickness increases from the left to the right, with an average value in the left, central, and right panels equal to about 6, 9, and 12 ML, respectively. See ref.

field. Anyway, comparison with results on thin films of NiO grown on Ag(001)[14] allows to conclude that the crystal field is responsible of little variations of the dichroic effect. The dichroic effects observed in going from $\theta = 90°$ *to* $\theta = 30°$ with **M** fixed, reveal then that the out-of-plane component of the average magnetic moments in NiO is much less than the in-plane one, reinforcing the conclusion that the anisotropy axis of NiO lies in the plane of the film. Similar deductions can be drawn from the data collected on the 20-ML-thick NiO film (not shown here), although in this sample the angular dependence of the absorption spectra appears to be less pronounced than for the 7-ML film, indicating a weaker anisotropy[13].

No significant temperature dependence of the spectra was noticed in the 180-300 K range, indicating that the Néel temperature of the NiO films is significantly higher than the room temperature.

The anisotropy of the NiO films can be better understood by analyzing the microscopy images acquired by means of a combined use of XMLD and PEEM, as shown in figure 2 (right side). The latter displays asymmetry images obtained at the Ni and Fe $L_2$ edges showing the magnetic contrast in a wedged NiO thin film grown over a Fe(001)-p(1×1)O substrate. In the asymmetry images dim areas are characterized by local magnetic moments aligning collinear to the light electric field, while bright areas indicate a perpendicular alignment, both for NiO and Fe. The NiO thickness increases from the left side to the right side of the figure and ranges between 6 and 12 ML. In the upper panels of Fig. 2 (right side), two different Fe magnetic domains are clearly visible for the substrate, separated by a domain wall parallel to the [110] direction. By knowing that our Fe(001) substrates have a remanent in-plane magnetization along one of its [100] directions (easy axes), we are able to conclude that the magnetization of the dark domain (topmost) is parallel to the [100] crystallographic axis, while the bright domain (lowermost) is magnetized parallel to the [010] axis. The observation of a magnetic contrast at the Ni $L_2$ edge proves that the NiO overlayer develops uniaxial anisotropy in the (001) plane. At low coverages (below about 9 ML, in good agreement with the results reported above) the NiO contrast is reversed with respect to Fe (panels a) and d) in the figure), indicating that the interfacial coupling between the Fe substrate magnetization and the NiO overlayer anisotropy axis is perpendicular, in concordance with what observed in the absorption experiments. Moving to higher NiO thickness values (central panels) the contrast disappears, in agreement with the reduction of the linear dichroism observed for thicker samples. For NiO thicknesses above about 9 ML (panels c) and f) in the figure), a magnetic contrast is retrieved on NiO and the observed magnetic contrast reveals that the type of interfacial coupling has turned from perpendicular to collinear.

## 5. Conclusions

XAS and XMCD measurements on the NiO/Fe(001) interface showed that reactive deposition of NiO on Fe promotes iron oxidation, leading to the formation of a buried layer of $Fe_3O_4$, which couples ferromagnetically to the iron substrate. This result is unexpectedly different from the case of the inverted Fe/NiO(001) interface where formation of antiferromagnetic FeO takes place. A sample annealing following the growth of the NiO/Fe interface is found to completely destroy the NiO overlayer, and the oxidation state of the resulting iron surface is very similar to FeO. These findings might have important consequences in F/AF/F devices where both types of interfaces would be present.

Furthermore, the NiO thin films show an in-plane magnetic anisotropy, which at low coverages causes a perpendicular alignement between the magnetic moments of NiO and Fe. This finding is once again radically different from what has been observed so far in F/NiO interfaces, where the coupling is collinear, indicating that also the magnetic structure of a F/AF interface can be heavily influenced by the growth sequence and conditions.

## Acknowledgments

The author warmly acknowledges all people at the Surface Physics group of the Department of Physics at Politecnico di Milano (Italy), the staff of the BACH beamline at the italian synchrotron ELETTRA and the staff of the PEEM-2 beamline at the ALS synchrotron in Berkeley (USA).

## References

1. V.E. Henrich and C. Cox, *The surface science of metal-oxides*, Cambridge University Press, Cambridge, 1994.
2. J. Nogués and I.K. Schuller, J. Magn. Magn. Mater. **192**, 203 (1999).
3. H. Matsuyama, C. Haginoya, and K. Koike, Phys. Rev. Lett. **85**, 646 (2000).
4. W. Zhu, L. Seve, R. Sears, *et al.*, Phys. Rev. Lett. **86**, 538 (2001).
5. H. Ohldag, T. J. Regan, J. Stöhr, *et al.*, Phys. Rev. Lett. **87**, 247201 (2001).
6. L. Duò, M. Portalupi, M. Marcon, *et al.*, Surf. Sci. **518**, 234 (2002).
7. A. Scholl, H. Ohldag, F. Nolting, *et al.*, Rev. Sci. Instrum. **73**, 1362 (2002).
8. R. Bertacco and F. Ciccacci, Phys. Rev. B **59**, 4207 (1999).
9. M. Finazzi, A. Brambilla, P. Biagioni, *et al.*, Phys. Rev. Lett. **97**, 097202 (2006).
10. T. J. Regan, H. Ohldag, C. Stamm, *et al.*, Phys. Rev. B **64**, 214422 (2001).
11. H.-J. Kim, J.-H. Park, E. Vescovo, Phys. Rev. B **61**, 15284 (2000).
12. M. Finazzi, A. Brambilla, L. Duò, *et al.*, Phys. Rev. B **70**, 235420 (2004).
13. M. Finazzi, M. Portalupi, A. Brambilla, *et al.*, Phys. Rev. B. **69**, 014410 (2004).
14. S. Altieri, M. Finazzi, H. H. Hsieh, *et al.*, Phys. Rev. Lett. **91**, 137201 (2003).

# NONLINEAR MAGNETO-OPTICAL PROBING OF MAGNETIC NANOSTRUCTURES: OBSERVATION OF NiO(111) GROWTH ON A Ni(001) SINGLE CRYSTAL

*V. K. Valev*[1,2,#], *A. Kirilyuk*[1] *and Th. Rasing*[1]
[1] *Institute for Molecules and Materials, Radboud University Nijmegen, Toernooiveld 1, 6525ED, Nijmegen, the Netherlands*
[2] *Institute for Nanoscale Physics and Chemistry, Katholieke Universiteit Leuven, Celestijnenlaan 200D, 3001, Heverlee, Belgium*

The magnetization induced effects in the nonlinear optical response of magnetic media lead to very strong and novel nonlinear magneto optical effects that appear to be very sensitive to magnetic interface properties, yielding very attractive applications for the study of magnetic thin films and multilayers. After a more general introduction, the application of nonlinear magneto-optics is illustrated with a more detailed study of the oxidation of Ni(001).

[#] V.K.Valev@fys.kuleuven.be

## 1. Introduction

Unique magnetic properties are revealed as we lower the dimensions of matter, from bulk materials to single monolayer interfaces and beyond. Among the many remarkable discoveries of the last fifty years are the exchange bias effect, giant magnetoresistance, tunneling magnetoresistance, interface anisotropy and interlayer exchange coupling. Together, these phenomena have revolutionized the field of data storage as they find applications in both random-access memories and hard drives. With the addition of the applications for new magnetic sensors, this revolution is in turn about to transform our daily lives as computers are no longer confined to desktops but are built in cars, stereos and even toasters.[1]

There exist several techniques (e.g. Spin Polarized Photoemission Spectroscopy, Spin Polarized Electron Energy Loss Spectroscopy, and Spin Polarized Low Energy Electron Diffraction to study the magnetic properties of clean surfaces[2]. Unfortunately, (polarized) electrons are difficult to use for studying buried interfaces due to their short mean free path. Since interfaces between thin metallic films are accessible by light, an optical technique has significant advantage.

Just as linear Magneto-Optics describes the interaction of polarized light with magnetic (or magnetized) materials, nonlinear magneto-optics describes their nonlinear optical response[3,4]. Physically, magneto-optical (MO) effects are related to the breaking of the time-reversal operation, leading to well-known phenomena as the Faraday effect and magnetic circular dichroism in transmission and the magneto-optical Kerr effect (MOKE) in reflection[5]. Similarly, nonlinear magneto-optical effects are related to a combination of time-reversal and space-inversion symmetry breaking (at least within the electric dipole approximation - see below), which makes them very interface specific.

Though the Faraday and Kerr effects are known for over a century, the field of nonlinear magneto-optics started much more recently with the prediction by Ru-Pin Pan and Shen[6] of magnetization induced contributions to the nonlinear optical response of magnetic surfaces and interfaces and the subsequent experimental observation of such effects by Reif *et al*[7] for an Fe surface and by Spierings et al[8] for Au/ Co interfaces. Actually, first predictions of possible nonlinear magneto-optical effects were already published as early as 1967 by Lajzerowicz and Vallade[9] for anti-ferromagnetic materials and by Akhmediev et al[10] for magnetic crystals; the "revival" and recent strong development of nonlinear magneto-optics is clearly related to the enormous interest in the study and applications of magnetic multilayers and nano-structures and by the development of solid state mode-locked femto second lasers that are particularly suitable for these kinds of studies.

Apart from the nonlinear optical equivalents of the Faraday and Kerr effects, the higher order optical response also leads to new nonlinear magneto-optical phenomena that have no equivalence in linear magneto-optics, such as transverse Faraday and Kerr effects that are linear in magnetization[11,12], the breaking of the equivalence between magnetization reversal and the change of light helicity[13] and the sensitivity for antiferromagnetic ordering[14]. In this paper we will focus on the applications of nonlinear optics for the studies of the magnetic properties of thin ferromagnetic films and interfaces.

**2. Theory**

An incident electromagnetic wave can induce a polarization in a medium that serves as a source for the transmitted and reflected light. This polarization **P** can be written in the electric dipole approximation as an expansion in powers of the electric field $\mathbf{E}(\omega)$ of the incident wave:

$$P_i = \chi_{ij}^{(1)} E_j\ (\omega) + \chi_{ijk}^{(2)} E_j\ (\omega) \quad E_k\ (\omega)\ +\ \dots \tag{1}$$

The tensor $\chi^{(1)}$ is the linear optical susceptibility, leading to the linear reflection and refraction of light. The second term describes nonlinear optical effects such as Second Harmonic Generation. In the presence of a magnetization **M** both $\chi^{(1)}$ and $\chi^{(2)}$ should be further expanded in powers of **M**:

$$\chi^{(1)} = \chi_0^{(1)} + \frac{\delta\chi^{(1)}}{\delta\mathbf{M}} . \mathbf{M} + \dots \tag{2a}$$

$$\chi^{(2)} = \chi_0^{(2)} + \frac{\delta\chi^{(2)}}{\delta\mathbf{M}} . \mathbf{M} + \dots \tag{2b}$$

One can think about the first term in both 2a and 2b as describing purely crystallographic effects while the second one only exists in the presence of magnetization.

In principle in both cases the magnetic contribution is described by a higher rank tensor; in practice these magnetic contributions can be written as effective $\chi^{(1)}$ and $\chi^{(2)}$ components respectively. For example, a nonzero magnetization along the $\hat{z}$-axis of an optically isotropic material yield nonzero components $\chi_{xy}^{(1)} = -\chi_{xy}^{(1)}$ that lead to linear magneto-optical effects that are proportional to the ratio $\chi_{xy}^{(1)} / \chi_{zz}^{(1)}$.

The crystallographic nonlinear optical tensor $\chi^{(2)}$ is only nonzero in media without inversion symmetry. Consequently, for isotropic media, $\chi^{(2)}$ only gives a contribution from surfaces and interfaces where the inversion symmetry is broken. The presence of a magnetization does not affect the inversion symmetry (as **M** is an axial vector) but does lower the symmetry of $\chi^{(2)}$ and consequently leads to new non-zero nonlinear susceptibility components. These elements can be derived in a straightforward way and are summarized in Table 1 for the standard longitudinal, transversal and polar magneto-optical configurations.[4]

| | $P_{in}$ | $S_{in}$ | |
|---|---|---|---|
| M‖*x* | xxz, zxx, zzz | zyy | $P_{out}$ |
| | **yxx, yzz** | **Yyy** | $S_{out}$ |
| M‖*y* | xzx, zxx, zzz, **xxx, xzz, zzx** | zyy, **xyy** | $P_{out}$ |
| | - | - | $S_{out}$ |
| M‖*z* | xxz, zxx, zzz | zyy | $P_{out}$ |
| | **yzx** | - | $S_{out}$ |

**Table 1.** The nonzero $\chi^{(2)}$ components for the longitudinal (M‖x), transversal (M‖y) and polar (M‖z) magneto-optical configurations (xz is the plane of incidence). For convenience, only the indices are indicated. The odd components are indicated in bold.

From this table, the effects of Magnetization induced Second Harmonic Generation (MSHG) can directly be understood. For example, in the longitudinal configuration, an s-polarized (parallel to y) input beam will give rise to one even component ($\chi^{(2)}_{zyy}$) and one odd component ($\chi^{(2)}_{yyy}$). Because the even term will yield a p-polarized and the odd term an s-polarized field respectively, the output polarization can be varied by changing the direction of **M**, yielding a nonlinear Kerr rotation $\theta^{(2)}_k \sim$ arctg $\left(\chi^{(2)}_{yyy} / \chi^{(2)}_{zyy}\right)$ in the same way as in linear magneto-optics, but with values for $\theta^{(2)}_k$ that are three orders of magnitude larger than the linear equivalent[15]. Similarly, it follows from Table 1 that in the transverse geometry both s- and p-polarized input beams will always generate a p-polarized output. Consequently, as in linear MOKE, there is no nonlinear Kerr rotation. However, in contrast to the linear case, there is a strong magnetization induced intensity change, that is <u>linear</u> in the magnetization.

Table 1 shows another important difference between the linear and the nonlinear magneto-optical effects: whereas in both cases the MO effects are proportional to a ratio between odd and even tensor components, in the linear case the odd components are small off-diagonal elements whereas in the nonlinear case the odd components are of a similar type as the even ones. This fact, that is also supported by calculations of $\chi^{(2)}$ based on spin-dependent band structure[16,17,18] is responsible for the very strong nonlinear magneto-optical effects that have been observed in various experiments and that typically are 3 orders of magnitude larger than the linear equivalents. A good introduction and overview of this field up to 1999 can be found in the book on Nonlinear Magneto Optics, edited by K. Benneman[3]. A more recent review can be found in ref.[19]

The relation between the nonlinear magneto-optical susceptibility and the spin dependent band structure can be written as:

$$\chi^{(2)}_{ijk}(2\omega) \propto \sum_{a,b,c} \frac{\langle a|i|c\rangle\langle c|j|b\rangle\langle b|k|a\rangle}{(2\hbar\omega - E_{ca} + i\hbar\Gamma_{ca})(\hbar\omega - E_{ba} + i\hbar\Gamma_{ba})} \tag{3}$$

Eq. (3) shows that spectroscopic MSHG experiments in principle allow to probe the spin-dependent surface or interface density of electronic states which is not only of fundamental interest but also highly relevant for the understanding of spin-dependent tunneling processes in magnetic tunnel junctions [20].

All the effects so far have been described within the electric dipole approximation only. An extension to include higher order contributions to the nonlinear magneto-optical response can be found in [21] whereas the application of Magnetization induced Sum Frequency generation can be found in [22].

## 3. Calculating the MSHG intensity

For an actual experimental configuration, the incoming electric field in Eq. 1 can be written as a sum of S and P polarization components, with unit vectors respectively $\hat{e}_S$ and $\hat{e}_P$ (see Fig. 1):

$$\mathbf{E}(\omega) = \hat{e}_S \cos(\psi) + \hat{e}_P \sin(\psi), \tag{4}$$

where $\psi$ is the angle between the incoming polarization and the S-direction of the polarizer. These unit vectors can be expressed in terms of the Cartesian coordinates for a coordinate system that we chose *on the sample*:

$$\begin{aligned} \hat{e}_S &= -\sin(\alpha)\hat{x} + \cos(\alpha)\hat{y} \\ \hat{e}_P &= \cos(\theta)\cos(\alpha)\hat{x} + \cos(\theta)\sin(\alpha)\hat{y} \pm \sin(\theta)\hat{z} \end{aligned}, \tag{5}$$

where $\theta$ is the angle of incidence and $\alpha$ is the angle between the plane of incidence and the x-direction. Replacing the unit vectors in Eq. 4 with the expressions in Eq. 5 we obtain the values of the local electric fields:

$$\begin{aligned} E_X^{IN} &= E_0(\cos(\theta)\cos(\alpha)\sin(\psi) - \sin(\alpha)\cos(\psi))\hat{x} \\ E_Y^{IN} &= E_0(\cos(\theta)\sin(\alpha)\sin(\psi) + \cos(\alpha)\cos(\psi))\hat{y}\,. \\ E_Z^{IN} &= E_0 \sin(\theta)\sin(\psi)\hat{z} \end{aligned} \tag{6}$$

The outgoing polarization can then be calculated by:

$$\begin{pmatrix} P_X^{OUT} \\ P_Y^{OUT} \\ P_Z^{OUT} \end{pmatrix} = \begin{pmatrix} \chi_{XXX} & \chi_{XYY} & \chi_{XZZ} & \chi_{XYZ} & \chi_{XXZ} & \chi_{XXY} \\ \chi_{YXX} & \chi_{YYY} & \chi_{YZZ} & \chi_{YYZ} & \chi_{YYZ} & \chi_{YXY} \\ \chi_{ZXX} & \chi_{ZYY} & \chi_{ZZZ} & \chi_{ZYZ} & \chi_{ZYZ} & \chi_{ZXY} \end{pmatrix} \cdot \begin{pmatrix} E_X^{IN\,2} \\ E_Y^{IN\,2} \\ E_Z^{IN\,2} \\ 2E_Y^{IN}E_Z^{IN} \\ 2E_X^{IN}E_Z^{IN} \\ 2E_X^{IN}E_Y^{IN} \end{pmatrix}, \tag{7}$$

where the susceptibility tensor represents the properties of the sample.

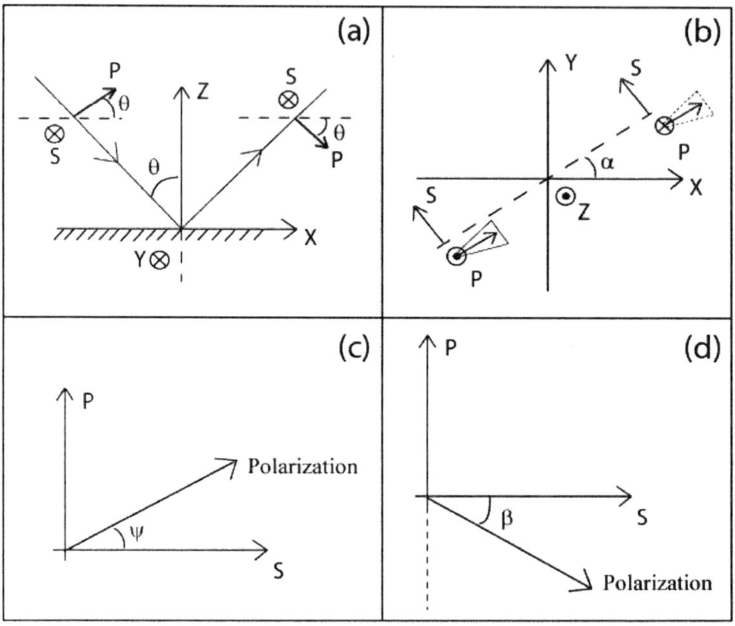


**Figure 1.** In (a), a schematic representation of the fundamental and the second harmonic beams relative to the sample as well as the incoming angle θ. In (b), α is the angle between the optical plane of incidence and the x-direction on the sample. In (c), ψ is the angle between the S-direction of the polarizer and the incoming polarization. In (d), β is the angle between the S-direction of the analyzer and the directions of outgoing second harmonic polarization.

Note that for simplicity we have neglected the Fresnel coefficients, which could also be (partially) included in the effective tensor components. The outgoing polarization is expressed in terms of S and P as:

$$\begin{pmatrix} P_S^{OUT} \\ P_P^{OUT} \end{pmatrix} = \begin{pmatrix} -\sin(\alpha) & \cos(\alpha) & 0 \\ \cos(\theta)\cos(\alpha) & \cos(\theta)\sin(\alpha) & -\sin(\theta) \end{pmatrix} \cdot \begin{pmatrix} P_X^{OUT} \\ P_Y^{OUT} \\ P_Z^{OUT} \end{pmatrix}. \quad (8)$$

The second harmonic polarization after the analyzer is then given by:

$$P_{Analyzer} = P_S^{OUT}\cos(\beta) - P_P^{OUT}\sin(\beta), \quad (9)$$

where $\beta$ is the angle between the analyzer and the S-direction.

Finally, the second harmonic intensity is calculated with:

$$I(2\omega) \propto \left|P_{Analyzer}\right|^2, \tag{10}$$

which gives the final second harmonic generation intensity.

## 4. Experimental considerations

The strong development of nonlinear magneto-optics in the past decade is also related to the development of solid state mode locked lasers that combine short pulse lengths with high repetition rates and allow the study of ultra thin magnetic films without destroying them. For most MSHG experiments nowadays, a Ti-sapphire laser (82 MHz x 100 fs pulses) tunable from 750-1100 nm but extendable to 400 nm –3 μm using a parametric amplifier, is used. After proper filtering, the generated specular harmonic light can be analyzed.

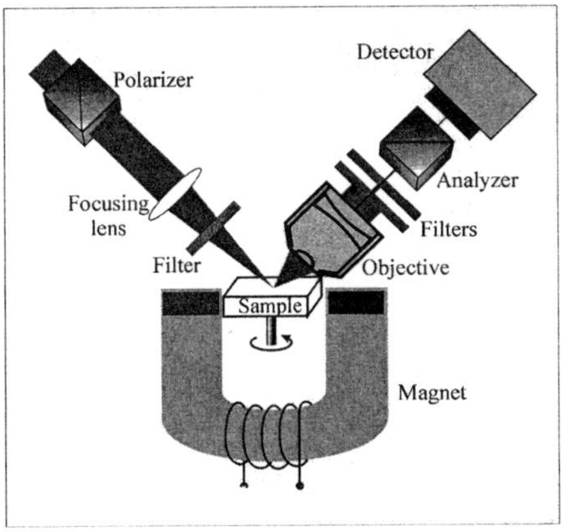


**Figure 2**. Schematic Configuration for nonlinear magneto-optical experiments. As detector, both photomultipliers or CCD camera's (for domain imaging) can be used. The input filter ensures that no SHG signal generated before the sample is detected, the filters at the output discriminate the SHG response from the reflected fundamental.

For each polarization combination, the total MSHG response from a magnetic material can be simplified by

$$\mathbf{P}(2\omega) = (\chi_{eff}^{even} \pm \chi_{eff}^{odd}(\pm\mathbf{M}))E_{\omega}^2 \tag{11}$$

where $\chi_{eff}^{even}$ and $\chi_{eff}^{odd}$ are effective tensor components that are even and odd in the magnetization and describe the crystallographic and magnetic contributions to the total response respectively. As both these contributions are complex quantities, the total (MSHG) signal is thus given by

$$I(2\omega) \sim |\chi_{eff}^{even}|^2 + |\chi_{eff}^{odd}|^2 \pm 2|\chi_{eff}^{odd}||\chi_{eff}^{odd}|\cos\Delta\Phi \tag{12}$$

where $\Delta\Phi$ is the phase difference between the two contributions. The importance of the latter is obvious: when $\Delta\Phi = \Pi/2$, the interference term is zero and changing the magnetization direction will have no effect on the total MSHG signal.

Though generally phase information is lost in intensity measurements, fortunately $\Delta\Phi$ can be measured quite easily in nonlinear optics by using interference techniques[23]. The latter can also be exploited in the case where there is only a purely odd response by adding a nonmagnetic reference signal, as it is the interference between even and odd terms that gives rise to the nonlinear MO effects.

Though MSHG signals give large relative MO effects, being a nonlinear optical technique the absolute intensities are rather small ($10$-$10^4$ photons/sec) but easily detectable with modern photon counting or charge coupled devices (CCD), though care should be taken to filter out the $2\omega$ signal versus the much stronger fundamental signal at $\omega$ (see Fig. 2).

Because of the simplicity of the experimental configuration coupled with the large effects, the transverse geometry is often used for experimental studies. One can then define a magnetic contrast or asymmetry as:

$$A = \frac{I(2\omega, +\mathbf{M}) - I(2\omega, -\mathbf{M})}{I(2\omega, +\mathbf{M}) + I(2\omega, -\mathbf{M})} \tag{13a}$$

$$= \frac{2|\chi_{eff}^{odd}|/|\chi_{eff}^{even}|}{1 + |\chi_{eff}^{odd}/\chi_{eff}^{odd}|^2}\cos\Delta\Phi \tag{13b}$$

Because A is normalized with respect to the total SHG intensity, it does not depend on the intensity or shape of the fundamental light pulses, nor on the spectral properties of optical components such as filters in the optical set up. Together with the already mentioned simplicity, this makes A useful parameter for quantitative investigations. One should however realize that the appearance of large effects that result from the large magnetic tensor components also means that, in contrast to most linear MO effects, the nonlinear effects are often not simply linearly proportional to the magnetization as directly follows from Eq's (12) and (13). This can for example strongly affect the shape of an MSHG loop[24,25].

## 5. Applications to magnetic surfaces and interfaces

A strong demonstration of the surface and interface sensitivity of the MSHG technique was given by Wieringa et al[12] in an in-situ MSHG and MOKE study of the Co/Cu(100) system (see Fig. 3 ). While the MOKE signal increased linearly with increasing Co thickness, as expected for a bulk sensitive probe, the MSHG contrast A first increased very rapidly but then stayed constant beyond 5 monolayers. The dashed lines in both sets of data are a result of model calculations and show a very good agreement with the experimental data. Further details on this study can be found in ref.[12]. These studies also showed that the nonlinear magneto-optical effects are quite sizeable: magnetic contrasts between signals with opposite magnetization directions of over 50% were observed. This indicates that the magnetization induced tensor elements are indeed of the same order of magnitude as the nonmagnetic ones.

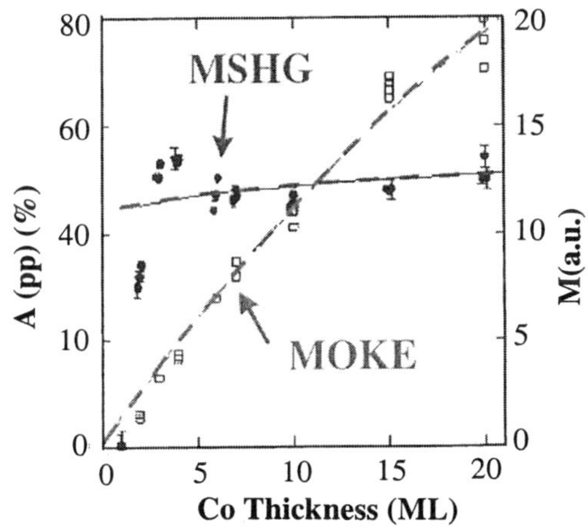


**Figure 3.** Comparison of MOKE (open symbols) and MSHG asymmetry (filled symbols) for a thin Co film on top of a Cu substrate.[12]

A demonstration of the spectroscopic possibilities is given in Figs. 4 and 5, showing the results of a phase sensitive spectroscopic MSHG investigation of a clean Ni(110) surface[26]. Fig. 4 shows how the magnetic asymmetry reaches a maxium around 2.7 eV that is completely quenched by exposing the surface to only 0.5 Langmuir of oxygen. This strong sensitivity to the surface condition confirms the surface sensitivity of the MSHG technique and suggests a surface state on Ni to be responsible for the observed resonances. This was further supported by a detailed analysis of these results, shown in Fig. 5. The data points in Fig. 5 are the experimentally obtained magnetic components $\chi_{eff}^{odd}$ that follow from the results in Fig. 4 by including the experimentally obtained phase information. For the phenomenological model calculations, the three states indicated in the bottom of Fig. 5 (the already mentioned surface state and the exchange split d-states) were used in Eq. (2) to calculate $\chi^{(2)}$. Excellent agreement is found. More recent first principle calculations by L. Calmels *et al* do support these findings[17,18]. These results show that the MSHG response is sensitive to transitions from exchange split d bands into empty surface states, and that the latter give rise to a strong enhancement of the MSHG response.

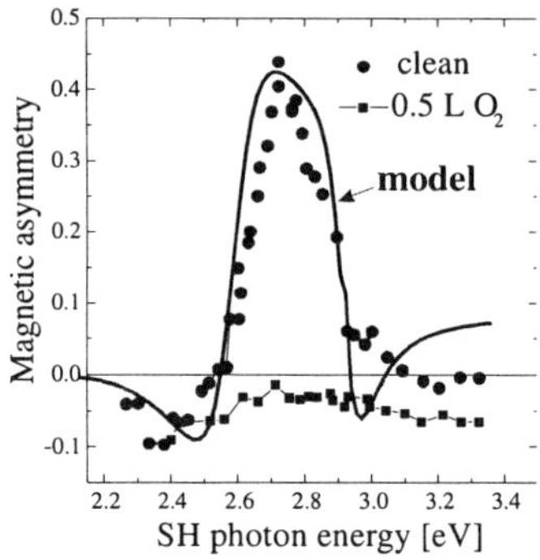


**Figure 4.** Magnetic asymmetry as a function of second harmonic photon energy as measured on a clean and oxidized Ni(110) surface. The solid line is a result of the calculation using 2 Lorentian lineshapes from Fig. 5.

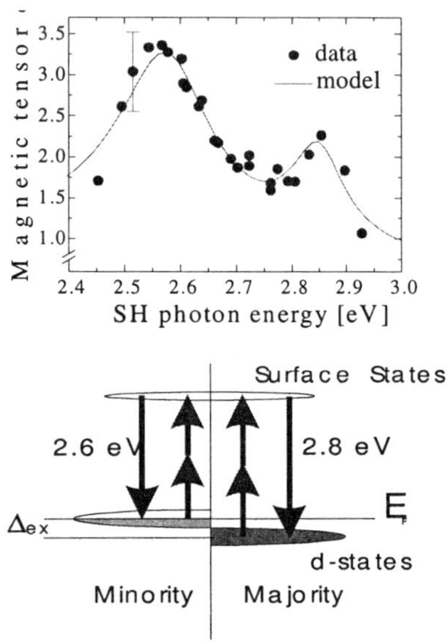


**Figure 5.** Effective magnetic component as derived from the measured intensity asymmetry and phase. The solid line gives a 2 Lorentzian fit using the states indicated in the bottom of the figure.

The growth of an oxide on single crystalline nickel also has some surprising results as far as the crystallographic symmetry is concerned. The next part shows very recent and still preliminary results on this interesting problem.

## 6. Symmetry considerations: observation of NiO(111) growth on a Ni(001) single crystal

As a second order nonlinear optical technique described by a third rank tensor, MSHG can also be exploited to probe other changes in symmetry, such as the appearance of steps on vicinal surfaces [27,28], magnetic moments at step edges[29] and the changes (phase transitions) in these properties.

Based on experimental results and on a MSHG intensity simulation, we present evidence that NiO grows in the (111) mode on top of Ni(001). This can have profound consequences on the magnetic properties of the system since, while NiO(001) contains layers of compensated AFM spins in the plane of the sample, the NiO(111) spins are uncompensated.

## 6.1. The oxidation of Ni

The interaction of oxygen with atomically clean nickel surfaces undoubtedly establishes one of the most studied surface reactions.[30] A comprehensive review about the oxygen-nickel reaction was given by Holloway,[31] and most relevant references can be found there. Virtually no surface sensitive technique has forgone to contribute to the understanding of this reaction, which, in fact, is a representative model system for the oxidation behavior of metal surfaces in general. In this sense, SHG is no exception and a detailed study of the nickel oxidation can be found in Ref. [32].

The reaction proceeds in three stages as a function of coverage. The first step is chemisorption. At room temperature, the oxygen forms ordered overlayers on the planes of nickel. In this chemisorption stage the oxygen atoms reside *above* favorable surface sites. Only the Ni(110) face shows a tendency to oxygen-induced surface reconstruction.[33]

The second stage is oxide nucleation to a depth of 2-3 atomic layers. A nucleation and growth model as proposed by Holloway and Hudson[34] has generally been accepted and has been verified with many techniques.[35] Various studies give deviating coverages at which nucleation first begins.[35] The nucleation rate is also temperature dependent,[36,37,38] and usually, the final thin NiO film is epitaxially related to the respective Ni face.

The third stage finally is a lateral growth to coalescence. Afterwards, a slow thickening process has been reported with different rate laws for different reaction temperatures.[35]

Of particular interest for this paper is that the epitaxial orientation of NiO on Ni(001) has been reported to follow not only (001) but also (111) and (7x7) patterns.[39,40,41]

## 6.2 Experimental details

The sample was a Ni(001) single crystal cleaned by Ar-sputtering in a UHV chamber and subsequently oxidized. SHG measurements of the clean Ni surface were performed *in situ* while those of the NiO were done *in air*. For the measurements we used a Ti-Sapphire laser at 800 nm with pulse width duration ~100 fs and a repetition rate of 82 MHz. The laser power was attenuated with a 1/20 chopper blade. The laser was focused to a spot with diameter of ~200 μm. The average laser power on the sample was 8 mW.

For the measurements *in situ*, because of the restrictions imposed by the design of the UHV chamber, the magnetic field was applied in the longitudinal configuration and the $S_{in}$-$45°_{out}$ polarization configuration was used so that the MSHG signal was sensitive to the longitudinal magnetization component. The sample holder was mounted on a UHV rotating motor.

For the measurements *in air*, the magnetic field was applied in the transverse configuration and the $P_{in}$-$P_{out}$ and $S_{in}$-$P_{out}$ polarization configurations were used so that the MSHG signal was sensitive to the transverse magnetization components only. For these experiments, the sample holder was mounted on a motorized and computer controlled rotational stage.

### 6.3 Results and discussion

In the absence of magnetism, the (001) surface of a cubic crystal lattice is described by a second-order susceptibility tensor, which we can write as:

$$\chi_{4-fold}^{cryst} = \begin{pmatrix} 0 & 0 & 0 & 0 & cr1 & 0 \\ 0 & 0 & 0 & cr1 & 0 & 0 \\ cr2 & cr2 & cr3 & 0 & 0 & 0 \end{pmatrix}, \qquad (14)$$

where $c1 \neq c2 \neq c3$. The SHG signal is then isotropic.

For a magnetic surface however, new tensor elements appear, which are *odd* in the magnetization, i.e. they change sign when the magnetization is reversed. If we chose to indicate these components with an "*a*", for a magnetic field applied in the direction of the optical plane of incidence, the second-order susceptibility tensor is:

$$\chi_{4-fold}^{magn} = \begin{pmatrix} 0 & 0 & 0 & 0 & cr1 & a2 \\ a2 & a1 & a3 & cr1 & 0 & 0 \\ cr2 & cr2 & cr3 & a3 & 0 & 0 \end{pmatrix}, \qquad (15)$$

where the "*cr*" components are *even* in the magnetization, i.e. they do not change sign upon magnetization reversal (note those are the same components as the crystallographic ones in Eq. 14).

For a magnetic field applied perpendicular to the optical plane of incidence, the corresponding tensor is:

$$\chi_{4-fold}^{magn} = \begin{pmatrix} a1 & a2 & a3 & 0 & cr1 & 0 \\ 0 & 0 & 0 & cr1 & 0 & a2 \\ cr2 & cr2 & cr3 & 0 & a3 & 0 \end{pmatrix}. \qquad (16)$$

We then see that for a magnetic field rotating around the sample, the tensor becomes:

$$\chi_{4-fold}^{magn} = \begin{pmatrix} a1\cos(\alpha) & a2\cos(\alpha) & a3\cos(\alpha) & 0 & cr1 & a2\sin(\alpha) \\ a2\sin(\alpha) & a1\sin(\alpha) & a3\sin(\alpha) & cr1 & 0 & a2\cos(\alpha) \\ cr2 & cr2 & cr3 & a3\sin(\alpha) & a3\cos(\alpha) & 0 \end{pmatrix}, \qquad (17)$$

where $\alpha$ is the angle between the x-axis of the crystallographic frame and the plane of incidence as described in Fig. 1. Here we chose that the direction corresponding to 0° is that for the transverse magnetic field geometry and at 90° is that for the longitudinal magnetic field geometry. In our experiments, while the sample is rotating the external magnetic field remains constant in the frame of the laboratory. Therefore, from the point of view of the sample, it is the field that is rotating (as well as the optical plane of incidence, the polarization vectors, etc.). Henceforth Eq. 17 applies again.

Manipulating the sample in a UHV chamber can be challenging since the various wires (such as the thermocouple or the sample heater) wind up during the rotation and might break. For this reason, we had to limit ourselves to a $\pm 90°$ rotation only, which is not a major setback since the studied surface is symmetric and therefore the complete rotation data can be extrapolated.

Significantly more problematic was the positioning of the sample surface perpendicularly to the optical plane of incidence. Under the combined weight of the magnet and the UHV motor, the sample holder mounting presented a small tilt and, as a consequence, during rotation the laser spot probed different regions of the sample. Small defects on the surface (such as scratches) could then constitute a source of noise in the signal. However, this problem could be reduced by measuring the magnetic asymmetry (or relative magnetic contrast) of the surface given by Eq. 13.

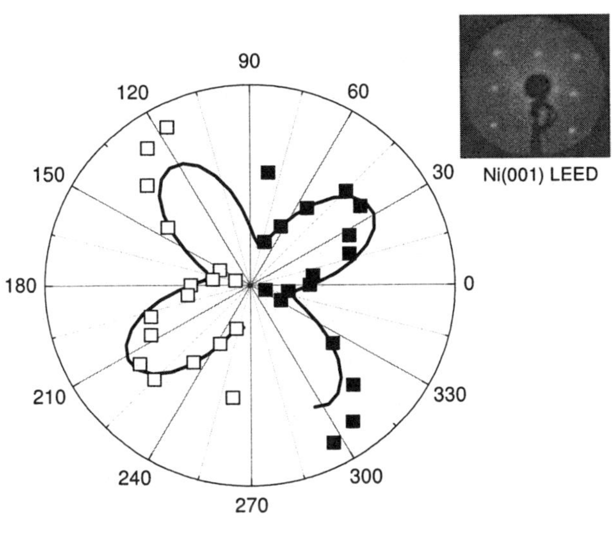


**Figure 6.** MSHG asymmetry of the clean Ni(001) as function of the azimuthal rotation of the sample in the UHV system for S-45° longitudinal geometry. The open squares on the graph were obtained by symmetry from the measured data (black squares). In inset, a LEED picture of the Ni(001) surface.

Fig. 6 shows the MSHG asymmetry curve obtained *in situ* from the clean Ni(001) surface. As expected, a 4-fold-like star is revealed with an angle of 90° between the peaks. The difference between the amplitude of the peaks can be explained by the sample surface tilt angle which introduces a slight additional 2-fold symmetry. In the inset a LEED picture shows the characteristic pattern of a cubic (001) lattice. The absence of oxygen on the surface was further confirmed by Auger Electron Spectroscopy.

The 4-fold fit was obtained according to the method described above. We substituted the susceptibility tensor of Eq.17 into Eq. 7 and we obtained the MSHG intensity from Eq. 10. The magnetic asymmetry was then calculated by Eq. 13. The parameters of the tensor components are presented in Table 2. Note again that for this calculation, we set our frame of reference on the sample.

| Ni(001) curve | cr1 | cr2 | cr3 | a1 | a2 | a3 | zero level |
|---|---|---|---|---|---|---|---|
| **$S_{IN}$-45°$_{OUT}$** | 0.8 | 1.94 | 2.94 | 1.6 | 0.6 | 2.4 | -0.51 |

**Table 2.** Values of the tensor elements and the corrected zero signal level, for the fit in Fig. 6.

Initially, the angle between the x-direction of the crystal lattice of the sample and the plane of incidence was -17.7° ($\varphi$ = 17.7). For an incoming beam polarized along the S-direction, the MSHG asymmetry can be equal to zero when, at a certain angle of the sample, the MSHG intensities proportional to the $\chi_{xyy}^{odd}$ and $\chi_{yyy}^{odd}$ components cancel each other. This angle of cancellation can be influenced by the presence of additional symmetries in the signal. In the fit, the zero signal level had to be corrected by subtracting 0.51 which can be partially explained by the presence of quadrupole contributions to the magnetization and the additional 2-fold anisotropy induced by the tilt angle between the sample surface and the optical plane of incidence.

It is clear that the fit in Fig. 6, although it reproduces the essential features of the Ni(001) 4-fold surface, is not perfect. Manipulating the sample inside the UHV presents difficulties that are in turn reflected in the observed results. In order to avoid these experimental limitations, we have oxidized the clean and crystalline Ni surface in the UHV chamber. Oxidation of the Ni(001) surface was performed *in situ* at room temperature until the oxide formation was complete (more that 10 000 L). Subsequently the MSHG anisotropy measurements were done *in air*, on a standard optical table.

Fig. 7 shows the MSHG intensity measured as a function of the azimuthal rotation for both magnetization directions and for P-P and S-P polarizer-analyzer configurations. The data show a strong magnetic contrast and an unexpected 3-fold symmetry. Between P-P and S-P the 3-fold stars exhibit opposite phase, therefore the maxima in intensity are not due to artifacts such as scratches on the surface. Another difference between these two configurations is the opposite magnetic contrast.

As the Ni(001) surface is isotropic, as far as the crystallographic contribution to the SHG is concerned, this 3-fold signal can only be induced by the presence of NiO. Assuming NiO(111) growth on the Ni(001) single crystal leads to an additional anisotropic nonlinear susceptibility component (see Eq. 15). All the above mentioned features are reproduced by the fits (solid curves in Fig. 7) corresponding to the presence of tensor elements from a 3-fold crystallographic for NiO(111) and a 4-fold magnetic symmetry group for Ni(001).

It is important to note that all four fits were calculated simultaneously with a single formula, which was again obtained according to the method described above. Furthermore, the parameters of the 4-fold magnetic tensor in Table 3 were constrained to those of the clean Ni(001) surface in Table 2 and are therefore identical to them.

The details are as follows: $\psi$ is the angle between the polarizer and the S-direction of polarization, it is along P for the blue and green curves ($\psi = \pi/2$ rad) and along S for the black and red curves ($\psi = 0$ rad). The noise level of the fit is set at 200 which is the minimum for the curves. The initial angle between the x-direction of the crystal lattice of the sample and the plane of incidence is 80° ($\varphi = 80$). The 3-fold crystallographic $\chi^{\mathrm{cr}}$ tensor is:

$$\chi_{3-fold}^{cryst} = \begin{pmatrix} c1 & -c1 & 0 & 0 & c2 & 0 \\ 0 & 0 & 0 & c2 & 0 & -c1 \\ c2 & c2 & c3 & 0 & 0 & 0 \end{pmatrix}. \tag{18}$$

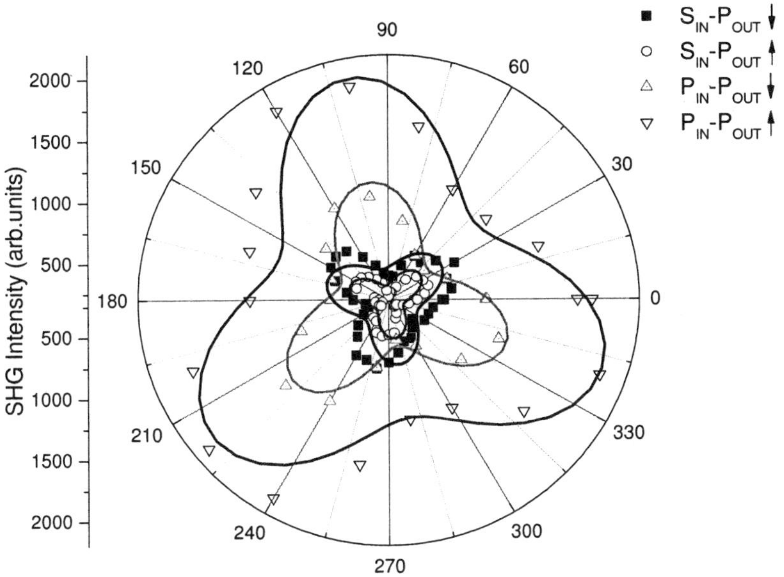


**Figure 7.** MSHG intensity as function of the azimuthal rotation of the sample for polarizer-analyzer configurations P-P and S-P for both directions of the magnetic field.

The entire fitting formula was multiplied by the amplitude constant "amp" which gives a correspondence between the units of the simulation and those of the experiment. It is 425 for both P-P curves and 38 for the S-P curves. The difference could partially be explained by the Fresnel coefficients which we did not take into account in our simulation. The 4-fold magnetic tensor components are again given by Eq. 17 with the same conventions.

In Table 3, it can be seen that the absolute values of the magnetic tensor elements a1, a2 and a3 are the same for all four fits, while their sign changes according to the changes of magnetic field. Furthermore, all the crystallographic tensor components and the experimental constants are the same.

| Curve | cr1 | cr2 | cr3 | a1 | a2 | a3 | c1 | c2 | c3 | amp |
|---|---|---|---|---|---|---|---|---|---|---|
| **$S_{IN}$-$P_{OUT}$ (down)** | 0.8 | 1.94 | 2.94 | **−1.6** | **−0.6** | **−2.4** | 0.8 | 1.94 | 2.94 | 425 |
| $S_{IN}$-$P_{OUT}$ (up) | 0.8 | 1.94 | 2.94 | **+1.6** | **+0.6** | **+2.4** | 0.8 | 1.94 | 2.94 | 425 |
| $P_{IN}$-$P_{OUT}$ (down) | 0.8 | 1.94 | 2.94 | **−1.6** | **−0.6** | **−2.4** | 0.8 | 1.94 | 2.94 | 38 |
| **$P_{IN}$-$P_{OUT}$ (up)** | 0.8 | 1.94 | 2.94 | **+1.6** | **+0.6** | **+2.4** | 0.8 | 1.94 | 2.94 | 38 |

**Table 3.** Values of the tensor elements and the signal amplitude, noise and phase constants for all the fits in Fig. 7.

## 7 Conclusion

Nonlinear Magneto-optics, or more specifically, Magnetization-induced Second Harmonic Generation has yielded an interesting and powerful new technique to investigate the magnetic structure of materials. The higher order optical response gives in particular the possibility of probing magnetic ordering in otherwise complicated or inaccessible structures like buried magnetic interfaces or anti-ferromagnetic materials and has been shown to lead to intrinsically new effects. Therefore, this new area of magneto-optics is not only interesting from a fundamental point of view but will also be quite important for applications such as the investigations of magnetic nanostructures and the spin dynamics in these systems. For recent reviews on this last topic see ref [42,43].

Examples on thin Co films and single crystalline Ni demonstrated the surface sensitivity, the large nonlinear magneto-optical effects and the spin dependent spectroscopy possibilities.

We have seen that the clean surface of Ni(001) single crystal has a 4-fold symmetry as revealed by LEED and the MSHG asymmetry. On the other hand, upon oxidation, we have observed a 3-fold MSHG pattern for different polarizer-analyzer combinations. These data can be very well fitted with calculations of the MSHG intensity including a 3-fold symmetry tensor. Using different experimental techniques, NiO(111) growing on Ni(001) has already been reported in the literature.[16,17,18] Therefore, we can conclude that we have observed NiO(111) growth on Ni(001) surface by means of MSHG for the first time.

The consequences of such an AFM/FM interface can be dramatic for exchange bias. Indeed, while a NiO(001)/Ni(001) is a compensated interface, NiO(111)/Ni(001) is uncompensated and this difference in interface spin order will very likely result in a different exchange interaction.

This work was supported in part by the EU project DYNAMICS, the Dutch organization for Fundamental Research on Matter (FOM) and the Dutch NanoNed program.

---

[1] Source: http://www.research.ibm.com/research/gmr.html
[2] See for example: Polarized Electrons in Surface Physics, R. Feder, ed. *(World Scientific, Singapore, 1985).*
[3] Benneman K H (ed.) *Nonlinear Optics in Metals*, Oxford University Press, Oxford 1998.
[4] Rasing, Th., Nonlinear magneto-optical probing of magnetic interfaces, *Applied Phys. B* **68** 477-484, 1999.
[5] Zvezdin K. and Kotov V.A., Modern Magneto-Optics and Magneto Optical Materials, studies in condensed matter physics. 1997 *(IOP Publishing Bristol and Philadelphia).*
[6] Pan R P, Wei H D, and Shen Y R 1989 Optical second-harmonic generation from magnetized surfaces. *Phys. Rev. B* **39,** 1229.
[7] Reif J, Zink J C, Schneider C M and Kirschner J 1991 effects of surface magnetism on optical 2nd harmonic-generation. *Phys. Rev. Lett.* **67** (20), 2878-2881.
[8] Spierings G, Koutsos V, Wierenga H A, Prins M W J, Abraham D and Rasing Th 1993 Optical second harmonic generation study of interface magnetism. *Surf. Sci.* **287**, 747-749; *J. of Magn. and Magn.* Mat. **121**, 109.
[9] Lajzerowicz J and Vallade M 1967 Génération du second harmonique dans les substances magnétiques ordonnées. *C.R. Acad.Sc Paris* **264**, 1819.
[10] Akhemiev N N, Borisov S B, Zvezdin A K, Lyubchanskii I L and Melikhov Y V, 1985 Nonlinear optical susceptibility of magnetically ordered crystals. *Sov. Phys. Solid State* **27** (4) 650-652.
[11] Pavlov V V, Pisarev R V, Kirilyuk A, Rasing Th 1997 Observation of a transversal nonlinear magneto-optical effect in thin magnetic garnet films. *Phys. Rev. Lett.* **78**, 2004-2007.
[12] Wierenga H A, Jong de W, Prins M W J, Rasing Th, Vollmer R, Kirilyuk A, Schwabe H, Kirschner J 1995 Interface magnetism and possible quantum well oscillations in ultrathin Co/Cu films observed by magnetization induced second harmonic generation. *Phys. Rev. Lett.* **74**, 1462-1465.
[13] Kirilyuk A, Pavlov V V, Pisarev R V and Rasing Th 2000a Asymmetry of second harmonic generation in magnetic thin films under circular optical excitation. *Phys. Rev. B*. **61**, R3796-3799.
[14] Fiebig M, Fröhlich D, Sluyterman vL G, Pisarev R V 1995 Domain topography of antiferromagnetic Cr2O3 by second-harmonic generation. *Appl. Phys. Lett.* **66**, 2906-2908.

[15] Koopmans B, Groot Koerkamp M, Rasing Th, Berg v d H 1995 Observation of large Kerr angles in the nonlinear optical response from magnetic multilayers. *Phys. Rev. Lett.* **74**, 3692-3695.
[16] Hübner W and Bennemann K H 1989 Nonlinear magneto-optical kerr effect on a nickel surface. *Phys. Rev. B* **40** (9), 5973-5979.
[17] L. Calmels, S. Crampin, J. E. Inglesfield, Th. Rasing: Surface states and second harmonic generation at the (110) Nickel surface from a first-principles calculation. *Surface Science* **482-485** (2001) pp. 1050-1055.
[18] Calmels-L; Inglesfield-JE; Arola-E; Crampin-S; Rasing-Th, Local-field effects on the near-surface and near-interface screened electric field in noble metals, *Phys. Rev B*.**64**, 12; 2001; pp.125416/1-8.
[19] A. Kirilyuk and Th. Rasing, J. Opt. Soc. Am. B, **22**, 148 (2005).
[20] Boer de P K, Wijs de G A and Groot de R A 1998 Reversed spin polarization at the Co(001)-HfO2(001) interface. *Phys. Rev. B* **58**, 15422-15425.
[21] Rasing Th. Nonlinear Magneto-Optical studies of ultrathin films and multilayers. In: Benneman K H (ed.) *Nonlinear Optics in Metals*, Oxford University Press, Chapter 3, pp 132-218, Oxford 1998.
[22] Kirilyuk A, Knippels G M H, Meer van der A F G, Renard S, Rasing Th, Heskamp L R and Lodder J C 2000 Observation of strong magnetic effects in visible-infrared sum frequence generation from magnetic structures. *Phys. Rev. B.* **62**, R783-R786.
[23] Stolle R, Veenstra K J, Manders F, Rasing Th, Berg v d H 1997 Breaking of time-reversal symmetry probed by optical second-harmonic generation. *Phys. Rev. B* **55**, R4925-4927.
[24] Veenstra K J 2000 Phase Sensitive Nonlinear Magneto-Optical Spectroscopy, PhD thesis University of Nijmegen, the Netherlands.
[25] V. K. Valev, M. Gruyters, A. Kirilyuk, Th. Rasing, Phys. Stat. Sol. (B) **242,** 3027 (2005).
[26] Veenstra K J, Petukhov A V, Jurdik E and Rasing Th 2000 Strong surface state effects in the nonlinear magneto-optical response of Ni(110). *Phys. Rev. Lett* **84**, 2002-2005.
[27] Leermakers R, Keen A M, Nguyen van Dau F, Rasing Th 2000 A Magneto-Optical Study of Co on Step Bunched Vicinal Substrates. *J. Magn. Soc. Japan,* **25** 275-278 2001.
[28] Frank A.R., Jorzick J, Rickart, M. Bauer, M. Fassbender, J. Demokritov S.O. and Hillebrands B, Scheib, M. Keen, A. Petukhov A., Kirilyuk A. Rasing, Th., Growth and magnetic properties of Fe films on vicinal to (001) substrates, *J. Appl. Phys.* **87** (2000) pp. 6092-6094.
[29] Jin Q J, Regensburger H, Vollmer R and Kirschner J 1998 Periodic Oscillations of the Surface Magnetization during the Growth of Co Films on Cu(001). *Phys. Rev. Lett.* **80**, 4056-4059.
[30] K. Wandelt, Surf. Sci. Rep. **2**, 1 (1982).
[31] P.H. Holloway J. Vacuum Sci. Technol. **18**, 653 (1981).
[32] K. J. Veenstra, *Phase Sensitive Nonlinear Magneto-Optical Spectroscopy*, Ph.D. thesis, Radboud University Nijmegen, the Netherlands (2000).
[33] J.F. van der Veen, R.G. Smeenk, R.M. Tromp and F.W. Saris Surface Sci. **79**, 212. (1979).

[34] P.H. Holloway and J.B. Hudson Surface Sci. **43**, 123 (1974).
[35] A.E. Berkowitz, K. Takano, J. Magn. Magn. Mater. **200**, 552 (1999).
[36] Dalmai-Imelik G, Bertolini J C and Rousseau J Surf. Sci. **63**, 67 (1977).
[37] P.H. Holloway and J.B. Hudson Surface Sci. **43**, 141 (1974).
[38] D.F. Mitchell, P.B. Sewell and M. Cohen Surface Sci. **61**, 355 (1976).
[39] W. -D. Wang, N. J. Wu and P. A. Thiel, J. Chem. Phys. **92** (3), 2025 (1989).
[40] O. L. Warren and P. A. Thiel, J. Chem. Phys. **100** (1), 659 (1993).
[41] E. Kopatzki and R. J. Behm, Phys. Rev. Lett. **74**, 1399 (1995).
[42] Th. Rasing, H. van den Berg, Th. Gerrits and J. Hohlfeld: Ultrafast magnetization and switching dynamics.In: "Spin Dynamics and Confined Magnetic Structures II, B. Hillebrands, K. Ounadjela (eds.), Springer-Verlag Berlin Heidelberg 2003. Topics Appl. Phys. **87** (2003) pp. 213-251.
[43] A.V. Kimel, C.D. Stanciu, P.A. Usachev, R.V. Pisarev, V.N. Gridnev, A. Kirilyuk, Th. Rasing:Phys. Rev. B. **74**} (2006) pp. 060403.

# PHOTOLUMINESCENCE UNDER MAGNETIC FIELD AND HYDROSTATIC PRESSURE IN $GaAs_{1-x}N_x$ FOR PROBING THE COMPOSITIONAL DEPENDENCE OF CARRIER EFFECTIVE MASS AND GYROMAGNETIC FACTOR

G. PETTINARI, F. MASIA, A. POLIMENI, M. FELICI, R. TROTTA, M. CAPIZZI
*CNISM and Dipartimento di Fisica, Università di Roma "La Sapienza", P.le A. Moro 2 Roma, 00185, Italy*

T. NIEBLING, H. GÜNTHER, P. J. KLAR, W. STOLZ
*Department of Physics and Material Sciences Center, Philipps-University, Renthof 5, Marburg, D-35032, Germany*

A. LINDSAY, E. P. O'REILLY
*Tyndall National Institute Cork, Ireland*

M. PICCIN, G. BAIS, S. RUBINI, F. MARTELLI, A. FRANCIOSI
*Laboratorio Nazionale TASC INFM-CNR, Trieste, Area Science Park 34012, Italy*

The influence of nitrogen cluster states (CS) on the conduction band (CB) structure of $GaAs_{1-x}N_x$ is probed by measuring the effective mass and gyromagnetic factor of carriers for a wide range of nitrogen concentrations (from $x < 0.01\%$ to $x = 0.7\%$). An unusual compositional dependence of these two important CB parameters is found. This behaviour is well reproduced by a modified **k·p** model taking into account the non-monotonic loss of Γ character of the CB minimum due to multiple crossings between the red-shifting conduction band edge and N cluster states. Analogously, sudden variations in the exciton mass can be externally induced by applying a hydrostatic pressure, which brings the upward moving CB edge into interaction with N states which at ambient pressure are resonant with the $GaAs_{1-x}N_x$ CB continuum.

## 1. Introduction

In dilute nitrides a small percentage (~1%) of nitrogen atoms is incorporated isovalently in a III-V crystal. Its introduction leads to the appearance of several N-cluster related states (CS) and to large modifications of the electronic properties of these compounds[1]. One of the most striking effects observed with increasing nitrogen concentration, $x$, is a large band gap decrease as a result of the mixing between different conduction band levels of the host crystal induced by the incorporation of single and multiple nitrogen complexes[2], which break the lattice translational symmetry[3]. In $GaAs_{1-x}N_x$, the pressure[4] and temperature coefficients[5] of the band gap are reduced sizably with respect to the N-free case, and the electron effective mass displays a rather unusual dependence on N concentration[6,7,8]. All these N-induced effects can be ascribed to a sizable distortion/mixing of the conduction band minima[3,9]. First, these effects of N were attributed simply to a crossing between a single N level and the conduction band (CB) edge[10].

## 2. Theoretical Model

The red-shift of the $GaAs_{1-x}N_x$ band gap energy with increasing N concentration was originally accounted for by a 2-levels band-anticrossing (BAC) model[11], where the wavefunction of the $GaAs_{1-x}N_x$ CB edge state can be obtained as $\psi_- = \alpha_c \psi_c + \alpha_N \psi_N$. Taking into account the interaction between the $\Gamma$ CB edge of GaAs, with wavefunction $\psi_c$, with a single nitrogen defect band state, with wavefunction $\psi_N$, which lies above the GaAs CB edge. Since the interaction between the N state and the GaAs CB edge increases with N composition, the $\Gamma$ character $f_\Gamma = |\alpha_c|^2$ of the CB edge state of $GaAs_{1-x}N_x$ smoothly decreases with $x$ [continuous line in Fig. 1]. This picture was deeply modified in Ref. 9 to take into account the interaction of the GaAs CB states with a distribution of nitrogen cluster states with energies both above and below the GaAs CB edge.

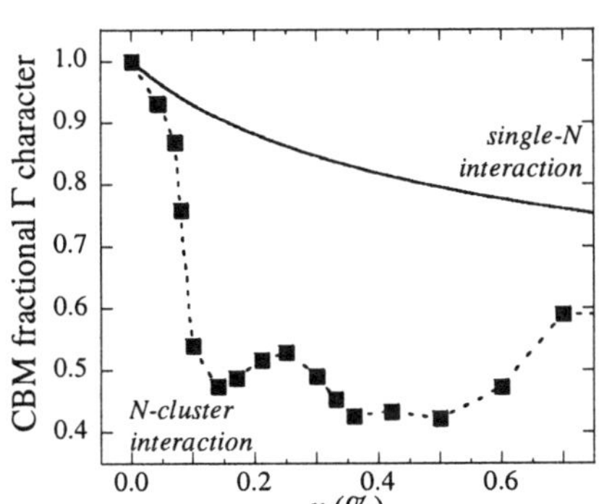


Figure 1. Dependence of CB minimum (CBM) fractional $\Gamma$ character on nitrogen concentration calculated both by BAC model and taking into account the interaction with N-related cluster states as in Ref. 9.

We model the effects of these clusters by placing L = 8000–10000 nitrogen atoms at random on the group V sites in a $Ga_MAs_{M-L}N_L$ supercell with 2M total atoms, and with composition $x$ = L/M. We use a tight-binding Hamiltonian to calculate the energies $\varepsilon_l$ and wavefunctions $\psi_{Nl}$ ($l$ = 1, ... , L) due to the interactions between the random distribution of nitrogen atoms considered. Then the CB minimum wavefunction can be obtained as a linear combination of isolated N states wavefunctions (LCINS). For increasing N concentration, the $GaAs_{1-x}N_x$ CB edge red-shifts and crosses the different N cluster levels. Depending both on the energy of the CB edge relative to the cluster state energies, $\varepsilon_l$, and also on the magnitude of the interaction for each state, $f_\Gamma$ will change in a *non-monotonic* way, as shown in Fig. 1 [full squares linked by dashed line].

$m_e$ and $g_e$ can be finally calculated by a 3 bands **k·p** model:

$$\frac{m_e(x)}{m_0}-1=\frac{P^2(x)}{3}\left(\frac{2}{E_g(x)}+\frac{1}{E_g(x)+\Delta_0}\right)$$
$$\frac{g_e(x)}{g_0}-1=-\frac{P^2(x)}{3}\left(\frac{1}{E_g(x)}-\frac{1}{E_g(x)+\Delta_0}\right) \tag{1}$$

where $E_g(x)$ is the band gap energy of $GaAs_{1-x}N_x$, $\Delta_0$ is the split-off energy, $m_0$ is the electron bare mass and $g_0 = 2$. The optical matrix element $P^2(x)$ measures the coupling between the CB minimum and the valence band maxima and it is proportional to the calculated $f_\Gamma$. Therefore, $P^2$ reflects the extent of interaction and admixing between the CB edge and the nitrogen CS.

## 3. Experiment and Results

We investigated about 20 $GaAs_{1-x}N_x$ samples having $x$ (or effective $x_{eff}$) ranging from 0% to 0.7% (obtained either by N incorporation during growth or by post-growth H irradiation[7], which passivates some of the nitrogen atoms leading to an effective $x_{eff}$). Magneto-photoluminescence was used to determine the electron effective mass, $m_e$, and the gyromagnetic factor, $g_e$, by measuring respectively the shift and the Zeeman splitting of the free-electron to neutral carbon luminescence band induced by a magnetic field ($B$ up to 12 T). The variation of $m_e$ and $g_e$ with $x$ is calculated by a modified **k·p** model taking into account hybridization effects between N cluster states (CS) and the CB edge[9]. A hydrostatic pressure ($P$ up to 10 kbar) was also applied to the samples in order to study the consequence of the alignment between the CB edge and nitrogen CS.

Figure 2 (a) shows the dependence of the electron mass on nitrogen concentration determined from the $B$-induced shift of the first Landau level for the electrons[7]. Squares and circles refer to as-grown and hydrogenated samples, respectively. After a first abrupt doubling ($m_e$ ~0.13 $m_0$) for $x$ ~ 0.1%, $m_e$ undergoes a second increase (~20%) for $x$ ~ 0.35%, finally it shows sizable fluctuations around $m_e$ ~0.14 $m_0$ for $0.4\% \leq x < 0.7\%$. A very similar qualitative behavior is found for the electron gyromagnetic factor[12], as shown in Fig. 2 (b). Indeed, $g_e$ exhibits a sign reversal for $x \geq 0.04\%$ and increases abruptly in a very narrow compositional window between $x = 0.04\%$ and $x = 0.1\%$. For $x > 0.1\%$, the electron gyromagnetic factor has a not well-defined behavior and fluctuates around 0.7 for the highest $x$ values.

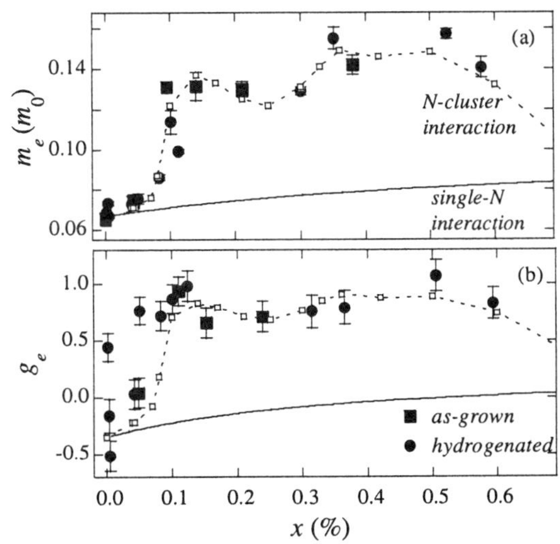


Figure 2. (a) Dependence of the electronic effective mass on N concentration (full symbols). (b) Same as (a) for the electron gyromagnetic factor. Theoretical estimation are calculated by Eqs. (1) using the fractional Γ character of Fig. 1, both in the framework of the BAC model (solid line) and by taking into account the interaction with the CS (open symbols) Dashed line is a guide for eye.

For $x$ ~0.1% and $x$ ~0.3–0.4% the CB minimum becomes resonant with N pairs and triplets, respectively, leading to a sizeable decrease in $f_\Gamma$ that according to Eqs. (1) produces an increase of electron effective mass and $g$-factor. Finally, notice the failure of the band-anticrossing model[10] to account for the experimental data [solid lines in Figures 2 (a) and (b)].

Quite interestingly, the degree of admixing between the free electron and the nitrogen CS can be finely tuned by application of a hydrostatic pressure. Indeed, due to the different rate of shifting of the N localized levels (~4 meV/kbar) and of the CB edge (~8–8.5 meV/kbar, respectively for $x$ = 0.21–0.1%), a crossing between these levels can be obtained. Figures 3 (a) and (b) show the dependence of the exciton effective mass $\mu_{exc}$ on applied pressure at $T$ = 90 K, respectively for $x$ = 0.21% and for $x$ = 0.1%, as determined by the diamagnetic shift of the band gap exciton[7]. Figures 3 (c) and (d) show the pressure dependence of the $T$ = 90 K energy position of the $GaAs_{1-x}N_x$ band gap (full circles and continuous line) and of nitrogen CS (open symbols and dotted lines) for $x$ = 0.21% and for $x$ = 0.1%, respectively. $\mu_{exc}$ displays a quite striking series of falls and rises, which can be traced back to the crossing between the CB edge and specific nitrogen CS resonant with the CB continuum, as highlighted by the vertical dashed lines in Fig. 3.

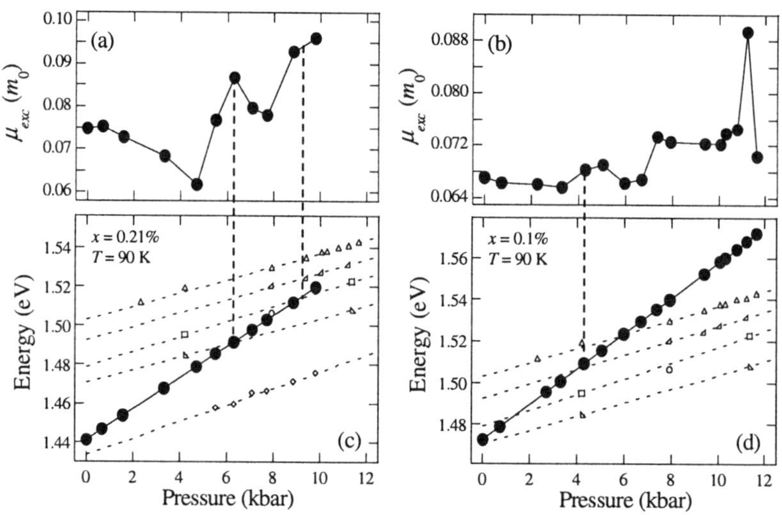


Figure 3. *Top*: Dependence on pressure of the exciton effective mass at $T$ = 90 K in $GaAs_{1-x}N_x$ with $x$ = 0.21% (a) and $x$ = 0.1% (b). *Bottom*: Dependence on pressure of the band gap energy (full circles) and CS (open symbols) in $GaAs_{1-x}N_x$ for $x$ = 0.21% (c) and $x$ = 0.1% (d). Vertical dashed lines indicate resonance condition between the CB and nitrogen CS.

## 4. Conclusions

We have shown that the nitrogen CS play a major role in determining the fundamental parameters of the CB structure of $GaAs_{1-x}N_x$. We were able to tune into or off resonance the CB edge and nitrogen-related levels both by varying the sample N concentration and by applying an external hydrostatic pressure. The ensuing variations of the CB effective mass and gyromagnetic factor provide a sensitive mean to investigate the physical properties of dilute nitrides.

## References


1. For a review see: *Physics and Applications of Dilute Nitrides*, edited by I. A. Buyanova and W. M. Chen (Taylor & Francis Books Inc., New York, 2004), and *Dilute Nitride Semiconductors*, edited by M. Henini (Elsevier, Oxford UK, 2005).
2. M. Bissiri, G. Baldassarri Höger von Högersthal, A. Polimeni, M. Capizzi, D. Gollub, M. Fischer, M. Reinhardt, and A. Forchel, *Phys. Rev. B* **66**, 33111 (2002).
3. P. R. C. Kent and A. Zunger, *Phys. Rev. B* **64**, 115208 (2001).
4. P. J. Klar, H. Grüning, W. Heimbrodt, J. Koch, F. Höhnsdorf, W. Stolz, P. M. A. Vicente, and J. Camassel, *Appl. Phys. Lett.* **76**, 3439 (2000).
5. A. Polimeni, M. Bissiri, A. Augieri, G. Baldassarri Höger von Högersthal, M. Capizzi, D. Gollub, M. Fischer, and A. Forchel, *Phys. Rev. B* **65**, 235325 (2002).
6. F. Masia, A. Polimeni, G. Baldassarri Höger von Högersthal, M. Bissiri, M. Capizzi, P. J. Klar, and W. Stolz, *Appl. Phys. Lett.* **82**, 4474 (2003).
7. A. Polimeni, G. Baldassarri Höger von Högersthal, F. Masia, A. Frova, M. Capizzi, S. Sanna, V. Fiorentini, P. J. Klar, and W. Stolz, *Phys. Rev. B* **69**, 41201(R) (2004).
8. F. Masia, G. Pettinari, A. Polimeni, M. Felici, A. Miriametro, M. Capizzi, A. Lindsay, S. B. Healy, E. P. O'Reilly, A. Cristofoli, G. Bais, M. Piccin, S. Rubini, F. Martelli, A. Franciosi, P. J. Klar, K. Volz, and W. Stolz, *Phys. Rev. B* **73**, 73201 (2006).
9. A. Lindsay and E. P. O'Reilly, *Phys. Rev. Lett.* **93**, 196402 (2004).
10. W. Shan, W. Walukiewicz, J. W. Ager III, E. E. Haller, J. F. Geisz, D. J. Friedmann, J. M. Olson and S. Kurtz, *J. Appl. Phys.* **86**, 2349 (1999).
11. J. Wu, W. Shan, W. Walukiewicz, K. M. Yu, J. W. Ager III, E. E. Haller, H. P. Xin, and C. W. Tu, *Phys. Rev. B* **64**, 085320 (2001).
12. G. Pettinari, F. Masia, A. Polimeni, M. Felici, A. Frova, M. Capizzi, A. Lindsay, E. P. O'Reilly, P. J. Klar, W. Stolz, G. Bais, M. Piccin, S. Rubini, F. Martelli, and A. Franciosi, *submitted to Phys. Rev. B.*

# PROBING THE DISPERSION OF SURFACE PHONONS BY LIGHT SCATTERING

Giorgio Benedek
Dipartimento di Scienza dei Materiali dell'Università di Milano-Bicocca, 20125 Milano, Italy, and Donostia International Physics Center, 20018 San Sebastian, Spain[1]
and
J. Peter Toennies
Max Planck Institut für Dynamik und Selbstorganisation,
Bunsenstrasse 10, 37073 Göttingen, Germany

Abstract

Although photons have, on the energy scale of visible light, a very small wavevector as compared to the Brillouin zone size of ordinary surfaces, various methods have been devised for measuring the dispersion curves of surface phonons with inelastic light scattering. More recently also inelastic X-ray scattering under grazing incidence and total reflection conditions has been successfully used to measure the dispersion curves of both surface and bulk phonons in the same sample. In this paper we briefly review these methods, with a comparison to inelastic helium atom scattering spectroscopy, arguing that measuring the dispersion curves of surface phonons may soon become a current branch of surface optical spectroscopy (*epioptics*).

## 1. Introduction

More than one hundred and twenty years ago Lord Rayleigh proved mathematically [1] that the long-wave component of earthquakes is due to a seismic wave traveling along the surface of the planet (Fig. 1). Rayleigh's theory describes the displacement amplitudes of waves which are localized near the surface of a semi-infinite elastic continuum. As shown in Fig. 2 they are polarized in a plane normal to the surface along the propagation direction (sagittal plane) [3]. In nature surface elastic waves actually exist over an extremely wide range of frequencies extending over about 23 orders of magnitude (Fig. 3) [4-6]. At the low frequency end are the seismic waves occurring on the surface of stars with wavelengths extending up to several thousands kilometers – four orders of magnitude longer than the ordinary earthquake waves.

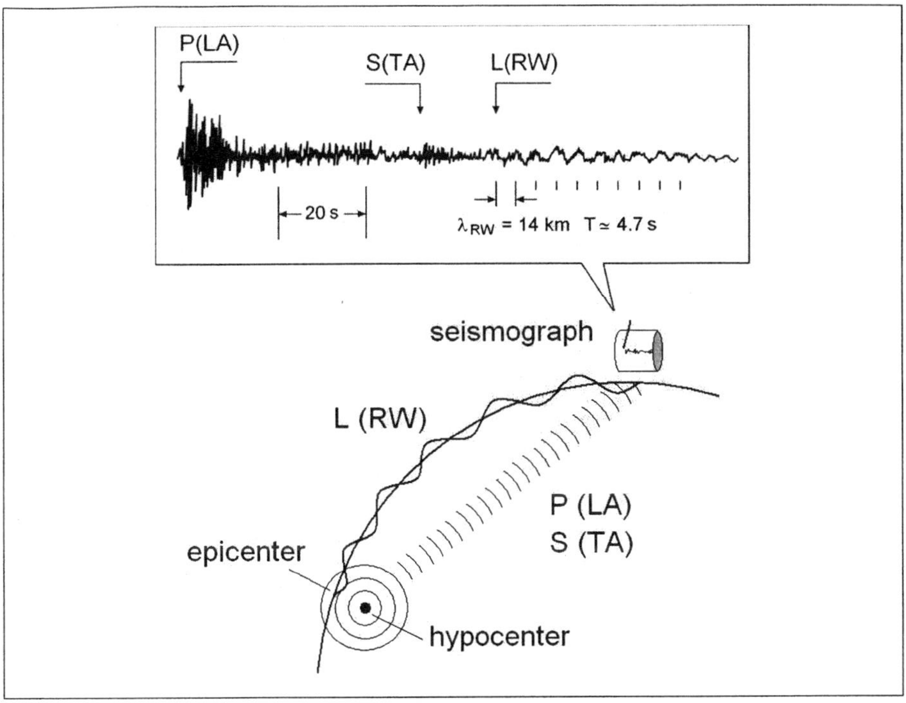


Fig. 1: (a) Seismographic pattern of the earthquake which hit south Italy in 1980 as detected at the Erice Seismic Network "Alberto Gabriele" of the Ettore Majorana Centre in Erice, 420 km away from the epicenter. The arrival times of the primary (P), secondary (S) and long-wave (L) components are indicated by arrows and correspond to the longitudinal acoustic (LA), transverse acoustic (TA) and Rayleigh waves (RW), respectively, in the notation of solid state physics. (b) LA and TA waves travel directly through the earth, and practically all frequencies of the spectrum reach the seismograph. On the contrary Rayleigh waves travel along the earth's surface, where the shorter wavelength components are strongly scattered by the earth surface asperities and only the long waves (about 14 km in the figure) reach the seismograph [2].

The other extreme of acoustic vibrations at short wavelengths has become the object of direct investigation through the development of interdigital transducers in piezoelectric materials, whose spectral range has been expanded over five decades from the audible region into the ultra- and hypersound regimes up to the THz domain. Recently it has become possible to visualize surface vibrations in the sub-GHz domain directly by means of stroboscopic X-ray diffraction. Figure 4 shows an example of the direct observation of a surface acoustical wave (SAW) generated in the piezoelectric material $LiNbO_3$ with a micrometric wavelength [7]. Interdigital transducers have found many applications and have opened up the technical development of SAW-based devices for applications in delay lines, opto-acoustic and signal processing systems [8-10]. Under the pressure of mobile telecommunication industry, the last two decades have witnessed the development of a new generation of SAW devices for multiband telephones [11-15], low-noise microwave oscillators [16], and frequency-modulated continuous-wave radars [17-18]. SAW devices combine a small size and integrability with operational frequencies in the GHz range, thus offering capabilities unattainable by other technologies.

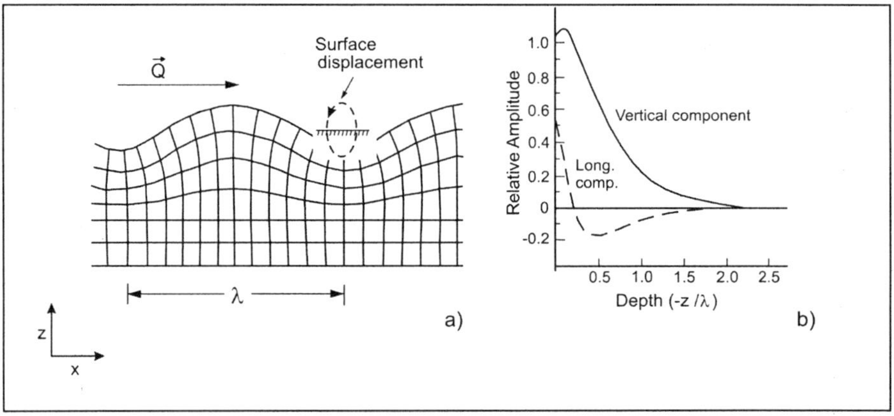


Fig. 2: (a) Schematic diagram showing the displacement pattern of a Rayleigh wave (RW) in the vicinity of the surface of a semi-infinite homogeneous elastic solid in the sagittal plane (from Farnell [1.8]). The displacement field has a dominant component normal to the surface. The smaller longitudinal component has a $\pi/2$ phase shift with respect to the normal one. Together they lead to a retrograde elliptical polarization in the sagittal plane. (b) Both components decay exponentially inside the solid with a decay length proportional to the RW wavelength $\lambda$, which is a characteristic of a *macroscopic* surface mode.

The extension of surface acoustic waves into the higher GHz and THz spectral regions is opening new perspectives in the broad field of sensors [19] as well as in acousto-optics and phononics [20]. The band structure and spectral gaps for surface acoustic waves have been recently demonstrated in piezoelectric phononic crystals [21,22] in a perfect analogy with to photonic crystals. The experimental study and applications of SAW's in the THz region, where wavelenghts reach the nanometric scale, have been fostered by the availability of nearly ideal defect-free surfaces. SAW's of a given wavelength are scattered by surface defects of comparable size and depth with an amplitude which is independent of the wavelength since the penetration depth scales with the wavelength (Fig. 4). Thus defects of any extension can be probed by SAW's of appropriate wavelength [7].

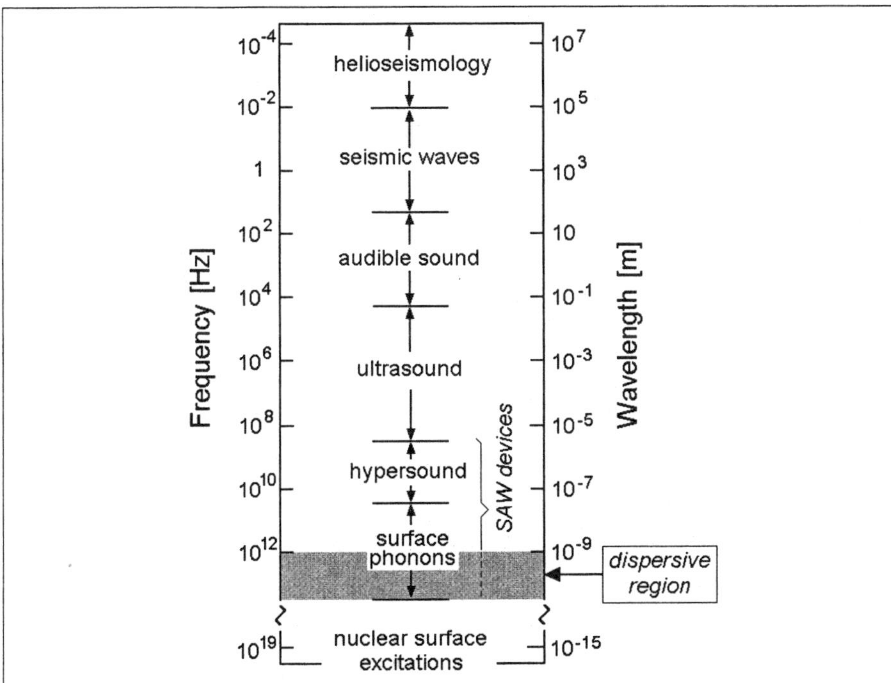


Fig. 3: Surface acoustic waves (SAW) cover a wide range of frequencies and wavelengths extending from $10^{8}$ m long waves on the surface of the sun [4] to $10^{-15}$ m for the vibrations on the surface of large nuclei [5]. The wavelength of ordinary Rayleigh waves vary over about 13 orders of magnitude, from the long-wave (L) component of earthquakes to the interatomic distances in crystal lattices. Over frequency regimes where the velocity of sound is constant, the frequencies are inversely proportional to the wavelength (adapted from Hess [6]). Applications of SAW devices extend nowadays to the THz domain, thus approaching the dispersive region where propagation velocities are no longer constant and quantization effects are important.

## 2. SAW physics with photons

Ultra-short laser pulses in the *ps* to the *fs* time scales and the related techniques of ultra-fast spectroscopy may be used not only to excite but also to monitor SAW's in the THz frequency domain by means of pump-probe experiments [6]. In one configuration two SAW's propagating in opposite directions are generated by the interference of two simultaneous *ps* laser pulses, and are detected through the diffraction of the probe laser beam. In this way Wright and Kawashima [23] have determined the surface-wave frequency spectrum of Mo and Cr films on a silica substrate. In another configuration a *ps* pulse of the pump laser generates a coherent SAW, which propagates along the surface and is monitored by a continuous probe laser beam through Michelson interferometry with a micrometric space resolution and frequencies up to 1 GHz [24]. More recently this pump-probe imaging technique of SAW ripples has allowed to directly measure the acoustic band structure and the Bloch harmonics in a $Cu/SiO_2$ phononic crystal [25].

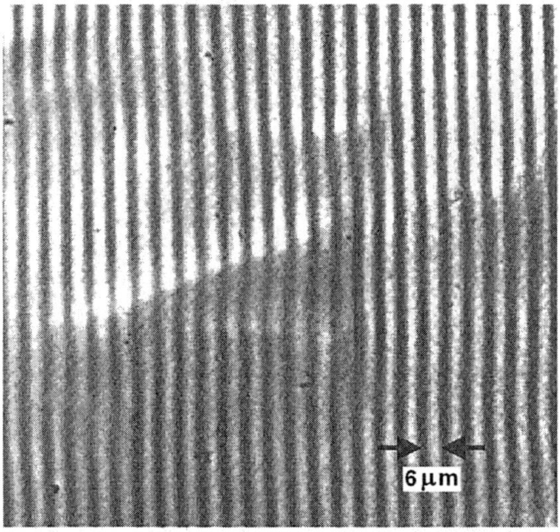


Fig. 4: A surface acoustic wave (SAW) of wavelength $\lambda$ = 6 μm and frequency $\nu$ = 0.58 GHz propagating along a $LiNbO_3$ as imaged by stroboscopic X-ray topography. The alterations of the wavefront profile and intensity reveal the presence of extended surface defects, e.g., dislocations (from D. Shilo and E. Zolotoyabko [7]).

The periodic acoustic deformation potential produced on a semiconductor surface by a SAW can separate electrons from holes and prevent recombination of excess carrier pairs [26]. The excess hole-electron pairs are created by a laser pulse at the surface of a ultra-thin InGaAs layer in a region which is crossed by a SAW. The SAW, which is produced by a radio-frequency generator, travels along the surface carrying a certain amount of excess electrons trapped at its minima and of excess holes trapped at its maxima. The charge carriers are transported along the surface with the SAW velocity until the SAW energy is absorbed by a transducer. At this point the hole-electron pairs suddenly recombine and the absorbed photons are re-emitted. The distance between adsorption and emission is 1 mm, corresponding to a delay of 350 ns. Many applications in composite devices, combining opto-acoustic with electronic functions, can be envisaged for recombination processes controlled by surface phonons.

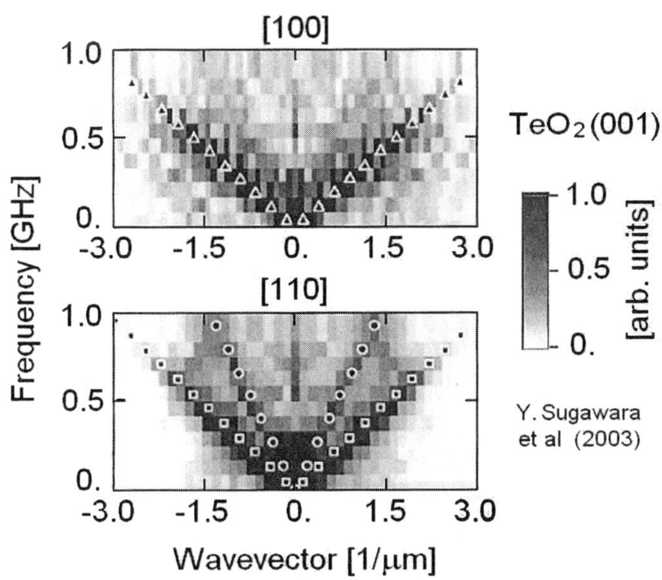


Fig. 5: The dispersion curves of the acoustic surface phonons in the GHz range along the symmetry directions of $TeO_2(001)$ as determined by pump-probe laser experiments by Sugawara et al [44] directly imaging the surface phonon displacement field (adapted from Y. Sugawara, O. B. Wright and O.Matsuda [44]).

A single-electron transport in a one-dimensional channel by means of high-frequency surface acoustic waves has been achieved by V. I. Talyanskii *et al* in a GaAs-AlGaAs heterostructure [27-28]. In this experiment the number of electrons trapped in the moving potential well of the SAW is controlled by the electron-electron repulsive potential and is well defined. According to the authors it may provide a tool for defining and producing a standard of electric current. The influence that SAW's have on the electric conductance through the opening of a ballistic quantum channel has been termed *acoustoconductance* [29].

Surface electrons can be trapped by a surface phonon associated with the local lattice deformation that the electron itself produces with its coulomb field (*self-trapping*). This special electron-phonon bound state is known as a *small polaron* [30] and results in a comparatively large electron effective mass. In this case the surface phonons are not injected from outside but are generated *in situ* by the excess electrons. The evidence for such a localization of surface excess electrons due to surface phonons has been obtained by Ge *et al* [31] in a state-of-the-art experiment based on the combination of angle-resolved two-photon photoemission (TPPE) and femtosecond laser techniques. In these experiments one pump photon lifts an electron from a (111) silver substrate into a surface image state of a bilayer of *n*-heptane molecules. The electron is localized in the direction normal to the surface plane by the overlayer-substrate interface but is allowed to freely move parallel to the surface with a parabolic (free-electron) dispersion. After a given pump-probe delay on the *fs* scale, a second (probe) photon extracts the electron with a defined kinetic energy, which depends on the emission angle. With such a time- and angle-resolved photoemission experiment it is possible to determine the image-state dispersion curve and even its evolution with time. A striking observation concerns the formation of the polaron state (which occurs on the time scale of a surface phonon period) and the consequent self-trapping of the image-state electron. The image-state electron energy as a function of parallel momentum is seen to evolve from the parabolic dispersion for free propagation (with an effective mass comparable to that of the free electron) for a vanishing delay, to a dispersionless branch corresponding to a self-trapped state after less than 2 *ps*. These experiments provide important information on the charge carrier dynamics in many low-dimensional systems including organic light-emitting diodes [32]. The interaction of SAW's with charge carriers in nanostructures under conditions of quantized transport has been investigated in many experiments on resonant tunneling [33], single-electron pumping through quantum dots [34-35], SAW single-electron interferometry [36] and SAW-driven light-emitting devices [37-38]. Even more, surface acoustic waves were demonstrated to provide a possible route to quantum computation [39-41].

It appears from these few examples that the dynamical interaction of quantized surface waves, i.e., surface phonons, with electrons and, via the electrons, with photon fields bears the promise for a large variety of new phenomena and applications. For example, femtosecond pump-probe experiments allow to visualize Raman-generated coherent optical phonon oscillations and to extract information on the electron-phonon interaction [42,43]. The latter experiments, however, are still concerning zone-center phonons. The measurement of the dispersion curves of the acoustic surface phonons in the GHz range with pump-probe laser experiments has been achieved a few years ago by Sugawara, Wright and Matsuda [44] through a direct imaging of the surface phonon displacement field along the symmetry directions of $TeO_2(001)$ (Fig. 5). The expected advantages from the extension of the above applications to the nanometric scale and the THz frequency range have promoted in the recent years new experiments on surface phonons in the high-frequency dispersive region by means of inelastic scattering spectroscopy with optical probes.

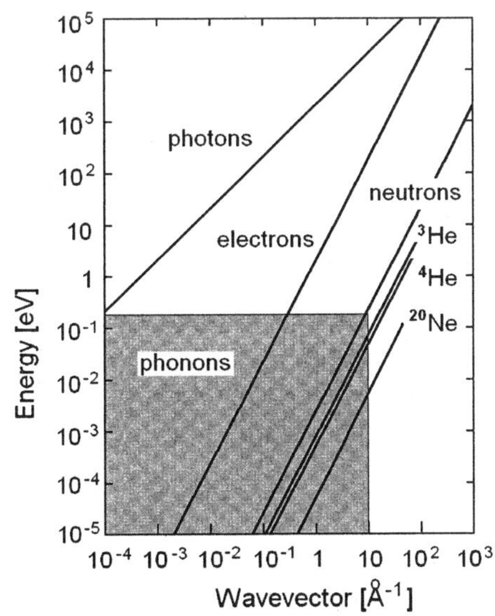


Fig. 6. Energy–wavevector relationship for various probe particles used for studying phonons. Only electrons, neutrons, helium and heavier atoms have energies and wavevectors which are comparable to those of phonons. Since neutrons can penetrate deep into the solid, they are not sensitive to the surface but ideal for measuring the dispersion curves of bulk phonons. Photons with wavevectors comparable to phonon wavevectors have energies a few orders of magnitude larger than phonon energies, falling in the X-ray domain, so that inelastic scattering spectroscopy requires a very high resolution.

### 3. Kinematics of Inelastic Scattering Spectroscopies: Photons versus Massive Particles

Inelastic scattering spectroscopy with either massive particles or photons is suitable to the study of surface phonon dispersion relations provided the incident energy and momentum are chosen in the appropriate range. Figure 6 illustrates the energy-wavector relationships for different probes which have been used to study surface phonons. Only electrons, neutrons, helium and heavier atoms have energies and wavevectors which are comparable to those of phonons. Since neutrons can penetrate deep into the solid, they are not sensitive to the surface unless measurements are performed on nano-crystalline powders, which makes them unsuitable to measure dispersion curves of surface phonons. Unlike massive particles, photons with wavevectors comparable to the size of a Brillouin zone have energies many orders of magnitude larger than phonon energies, which fall in the X-ray domain. Thus it is possible in principle to measure the surface phonon dispersion curves with X-ray inelastic scattering spectroscopy, but this requires a very high resolution of the order of at least $E/\Delta E \sim 10^7$. This resolution, while hardly achievable with massive particles (save $^3$He spin-echo techniques [45]), is well within the reach of photon spectroscopy. On the other hand, X-rays pose the same problem of neutrons – a large penetration into the solid, i.e., a weak sensitivity to surface phonons. This difficulty can be overcome, however, with a grazing incidence and/or under total-reflection conditions [46], which restricts the wavevector conservation to the surface components.

As appears in Fig. 6, for a *planar scattering* configuration the energy transfer $\Delta E = -\hbar\omega(\mathbf{Q})$, with $\omega(\mathbf{Q})$ the phonon frequency and $\mathbf{Q}$ the phonon wavevector, is related to the parallel momentum transfer $\Delta\mathbf{K} = -\mathbf{Q} + \mathbf{G}$ (with G a surface reciprocal lattice vector) by the linear relationship

$$\frac{\Delta E}{E_i} + 1 = \frac{\sin\vartheta_i}{\sin\vartheta_f}\left(1 + \frac{\Delta K}{K_i}\right) \qquad \text{for photons,} \qquad (1)$$

or by the quadratic relationship

$$\sqrt{\frac{\Delta E}{E_i} + 1} = \frac{\sin\vartheta_i}{\sin\vartheta_f}\left(1 + \frac{\Delta K}{K_i}\right) \qquad \text{for massive particles,} \qquad (2)$$

where $E_i$, $K_i$ and $\vartheta_i$ are the energy, the parallel wavevector component and the incidence angle of the incoming probe, respectively, $\vartheta_f$ is the final angle, and $\Delta K$ replaces $\Delta\mathbf{K}$ for the one-dimensional planar-scattering case. For a fixed scattering geometry, i.e., for given $\vartheta_i$ and $\vartheta_f$, these two equations are represented in the $(\Delta K, \Delta E)$-plane by the so-called *scan curves*, which are either a straight line (photons, Eq. (1)) or a parabola (massive particles, Eq. (2)).

In inelastic X-ray scattering (IXS) experiments a *non-planar scattering* configuration turns out to be more convenient [46]. The complex geometry for non-planar scattering from a crystalline surface is illustrated in Fig. 7 and by its comprehensive caption. In this general case the relationship between $\Delta E$ and $\Delta \mathbf{K}$ is expressed by two equations

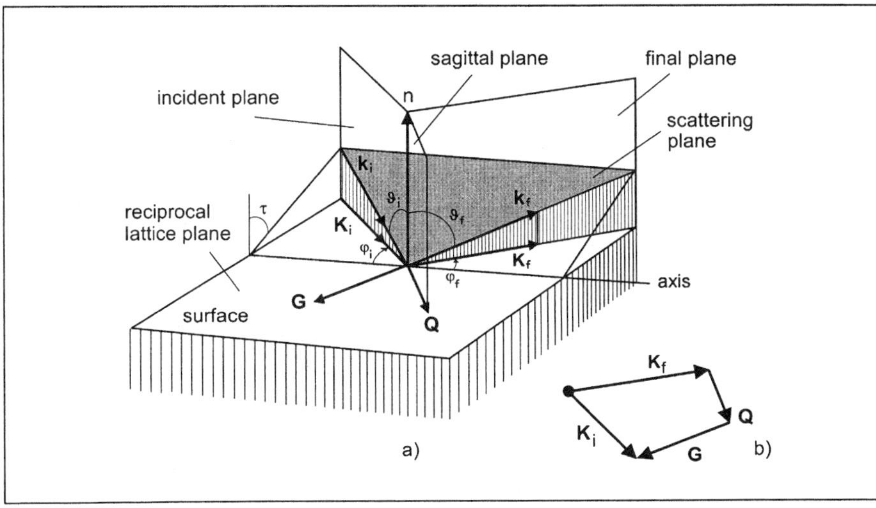


Fig. 7. (a) Schematic diagram where a monoenergetic beam of photons or particles is scattered from a crystal surface in a general *non-planar scattering* geometry. The particles are scattered from the *incident plane*, defined by the incident wavevector $\mathbf{k}_i$ and the surface normal $\mathbf{n}$, into the *final plane*, defined by the final wavevector $\mathbf{k}_f$ and the surface normal. The scattering is said to be non-planar when the incidence and final planes do not coincide. The wavevectors $\mathbf{k}_i$ and $\mathbf{k}_f$ define the *scattering plane*, which for non-planar scattering is not normal to the surface but forms an angle $\tau$ (*tilt angle*) with $\mathbf{n}$. The intersection of the scattering plane with the surface defines the scattering plane *axis*. The incidence and final particle directions are defined by the incidence and final angles, $\vartheta_i$ and $\vartheta_f$, and the azimuthal angles $\varphi_i$ and $\varphi_f$, respectively. For planar scattering $\varphi_i = \varphi_f = 0$. (b) In an inelastic process in which a single surface phonon is annihilated or created the parallel component $\mathbf{K}_i$ of the incident wavevector equals the sum of the parallel component $\mathbf{K}_f$ of the final wavevector, plus the phonon wavevector $\mathbf{Q}$ and, in Umklapp processes, a reciprocal surface lattice vector $\mathbf{G}$. All these vectors lie in the reciprocal surface lattice plane. Surface phonon polarizations are referred to the *sagittal plane*, defined by $\mathbf{Q}$ and $\mathbf{n}$. In the *planar scattering* geometry the incidence and final planes coincide ($\varphi_i = \varphi_f = 0$), and the scattering plane is normal to the surface. For planar inelastic scattering the sum $\mathbf{Q} + \mathbf{G}$ must also lie in the scattering plane, and for a scattering plane along a high symmetry direction of the surface, $\mathbf{Q}$ and $\mathbf{G}$ must both belong to this plane, which coincides with the sagittal plane.

$$\frac{\Delta E}{E_i}+1=\frac{\sin\vartheta_i}{\sin\vartheta_f}\left(\frac{\cos\varphi_i}{\cos\varphi_f}+\frac{\Delta K_{\parallel}}{K_i\cos\varphi_f}\right)=\frac{\cos\vartheta_i}{\cos\vartheta_f}\left(1+\frac{\Delta K_{\perp}}{K_i\sin\varphi_i}\right),\quad \text{for photons} \qquad (3)$$

where $\Delta K_{\parallel}$ and $\Delta K_{\perp}$ are the components of $\Delta\mathbf{K}$ parallel and perpendicular to the scattering plane axis, and $\varphi_i$ and $\varphi_f$ are the initial and final azimuth angles, respectively. A similar equation, with the left-hand member replaced by its square root, holds for massive particles. For a fixed scattering geometry, the two equations (3) describe in the ($\Delta K$, $\Delta E$) plane two scan curves, actually straight lines for photons, with respect to the components $\Delta K_{\parallel}$ and $\Delta K_{\perp}$ of the parallel wavevector transfer. Thus for each value of the energy transfer $\Delta E = -\hbar\omega(\mathbf{Q})$ the scan curves give simultaneously the two components $\Delta K_{\parallel}$ and $\Delta K_{\perp}$ and therefore $\mathbf{Q}$, after subtraction of a suitable $\mathbf{G}$ vector so as to bring $\mathbf{Q}$ into the first Brillouin zone

## 4. Overcoming the Problem of Small Photon Momentum

In the IXS spectroscopy the problem of the small photon momentum has been overcome by using incident energies in the high keV range, at the expense of energy resolution. Other methods have been devised, however, in order to measure the dispersion relations of surface phonons by using visible or UV photon probes, so as to exploit the superior energy resolution available in this range.

In one method (*grating light scattering*) inelastic light scattering can be used for a surface decorated with a periodic array of atomic rows perpendicular to the surface direction along which the phonon dispersion curves are to be investigated. The configuration is schematically shown in Fig. 8. The grating of parallel atomic rows, shown in a vertical section (a) and from above (b), has a period $A$ which is a multiple of the surface spacing $a$ and yields a multiple folding of the Brillouin zone (c) from the original size $\pi/a$ to the reduced size $\pi/A$. The dispersion curve of a surface acoustic phonon (e.g., the Rayleigh wave) is then folded many times into the reduced zone, and the points of the original curve (•) which are folded into the zone center (✦) can be measured by inelastic light scattering. If the dynamical perturbation of the grating is weak the splitting occurring at the folding points is sufficiently small so that single peaks are measured in correspondence with the original surface modes of the clean surface. In this way Dutcher *et al* [47] and Giovannini *et al* [48] have investigated the surface acoustic phonon dispersion curves of the reconstructed Si(001) p(2x1) surface with surface Brillouin scattering.

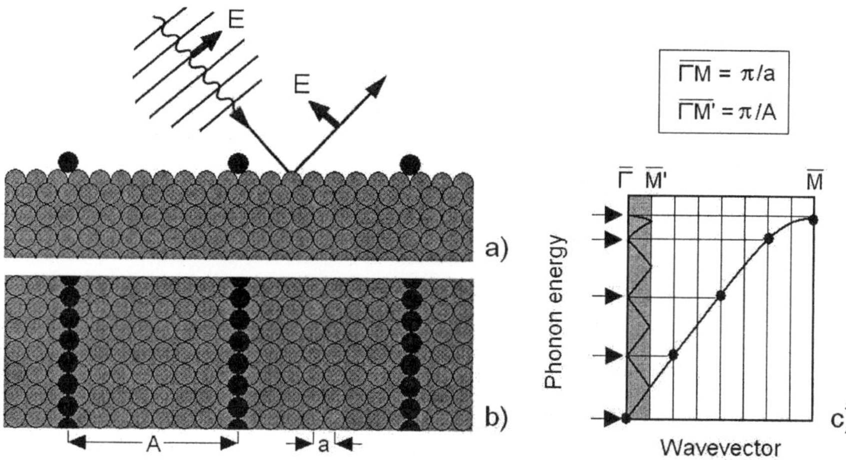


Fig. 8. In the grating light scattering method a grating of parallel atomic rows, shown in a vertical section (a) and from above (b), produces a folding of the Brillouin zone (c) from the original size $\pi/a$ to the reduced size $\pi/A$. The dispersion curve of a surface acoustic phonon is then folded many times into the reduced zone and the points of the original curve (●) which are folded into the zone center (➔) can be measured by inelastic light scattering. With respect to atom or electron scattering, light scattering has the advantage of the additional degree of freedom associated with the electric field polarization (E), which may allow to select the polarization of the excited phonons.

With respect to atom or electron scattering, light scattering has the advantage of the additional degree of freedom associated with polarization, which may allow selecting the polarization of the excited phonons. In massive particle scattering such a selection may be obtained with molecules with a $J \neq 0$ rotational ground state (e.g., ortho-$H_2$), through the so inelastic rotation-flip scattering [49] or with some complicated non-planar scattering configuration [50].

Another method to measure the phonon dispersion curves by inelastic light scattering is the double resonance Raman scattering (DRRS) [51,52]. The double-resonance condition may occur in multi-valley systems, such as, e.g., graphite, which has six equivalent Fermi points at the corners $\overline{K}, \overline{K}'$ of the Brillouin zone (Fig. 11), with the equivalent $\overline{K}(\overline{K}')$ points connected by a **G** vector. Figure 9 illustrates the kinematics of both the one-phonon defect-induced (a) and the two-phonon (b) version of DRRS. In the first case (a) a photon induces a resonant vertical transition of an electron near a $\overline{K}$ point followed by an inelastic transfer to a valley $\overline{K}'$ with emission of a phonon Q; then the electron resonantly recombines with the hole left behind at $\bar{K}$ exploiting a defect symmetry breaking which provides the missing momentum and the photon is re-emitted at a smaller energy. In the second case (b) the recombination occurs via a resonant vertical transition followed by the emission of another phonon identical to the first one but in the opposite direction. In both cases the phonon wavevector is easily determined from the knowledge of the band structure and the dispersion curve can be, at least in a substantial portion, by simply changing the incident photon energy.

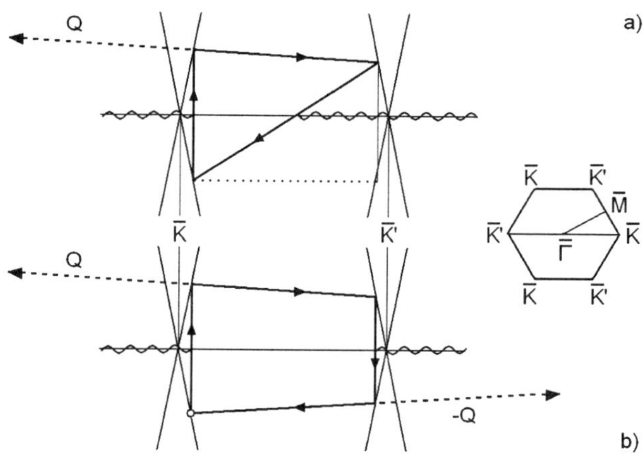


Fig. 9. The kinematics of one-phonon defect-induced (a) and two-phonon (b) double-resonance Raman scattering (DRRS) in the case of graphite. In the first process (a) a photon induces a resonant vertical transition of an electron between the valence and the conduction band near a $\bar{K}$ point of the planar Brillouin zone (inset), followed by an inelastic transfer to a valley $\bar{K}'$ with emission of a phonon Q; then the electron resonantly recombines with the hole left behind at $\bar{K}$ exploiting a defect symmetry breaking which provides the missing momentum and a photon is re-emitted at a smaller energy. In the second process (b) the recombination occurs via a resonant vertical transition followed by the emission of another phonon identical to the first one but in the opposite direction.

Figure 10 shows a comparison of the dispersion curves of graphite measured in this way with a bond-charge model calculation of the (0001) surface phonons [53] and with a more recent ab-initio calculation of the bulk phonon dispersion curves [51]. It appears that for a weakly-bound layer structure like graphite there is practically no difference between bulk and surface modes. Even in the extreme acoustic limit the Rayleigh wave branch is very close to the lower edge of the bulk acoustic band and also the surface optical modes can hardly be resolved from the very narrow bulk optical bands. Both BCM and ab-initio calculations are in substantial good agreement with experiment, but there is one important feature revealed by DRRS which is nicely reproduced by the ab-initio calculation and (obviously) not by the bond-charge model calculation: the Kohn anomaly (KA) predicted to occur around $\overline{\Gamma}$ point (and also at the $\overline{K}, \overline{K}'$ points) and consisting in a dip of the LO branch. The observation of this feature demonstrates the good resolution that the optical probe can afford with respect to other scattering probes in the high-frequency regions of the phonon spectra.

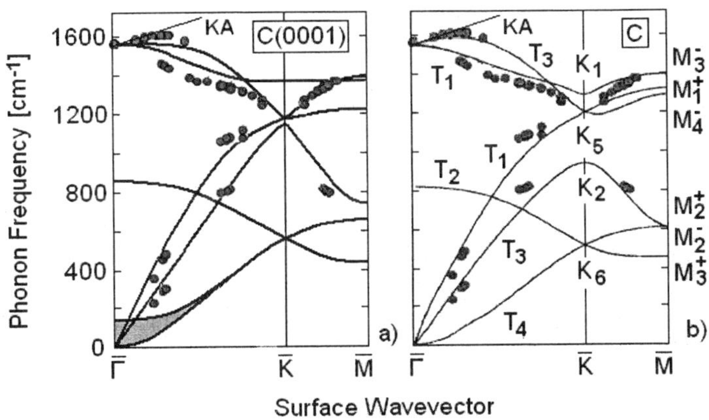


Fig. 10. A comparison of the dispersion curves of graphite measured with DRRS by Maultzsch *et al* [51,52] with (a) a bond-charge model calculation of the (0001) surface phonons [53] and (b) a more recent ab-initio calculation of the bulk phonon dispersion curves [51].

## 5. Inelastic X-ray scattering (IXS)

Also X-ray inelastic scattering spectroscopy has achieved in recent years a sufficient resolution (better than $E/\Delta E \sim 10^7$) and sensitivity to allow for the study of surface phonon dispersion curves [46,54]. As appears in Fig. 6, X rays of about 20 keV, as used in Ref. [54], have a wavevector of about 20 Å$^{-1}$. Thus IXS spectroscopy is approaching the required resolutions of $\Delta K \sim 0.1$ Å$^{-1}$ and $\Delta E \sim 1$ meV for a surface phonon spectroscopy competitive with the more conventional techniques with massive probes. Moreover IXS spectroscopy has the added values of exploiting photon polarization and the possibility of varying the penetration depth. In order to reduce the penetration of

X rays into the solid bulk and to use IXS as a genuine surface probe grazing incidence must be used with the grazing incidence angle kept below the total reflection critical angle $\alpha_c$. However, with grazing angle slightly above $\alpha_c$, penetrations of the order of 100 nm may be achieved [46,54] so as to measure essentially the bulk phonon structure.

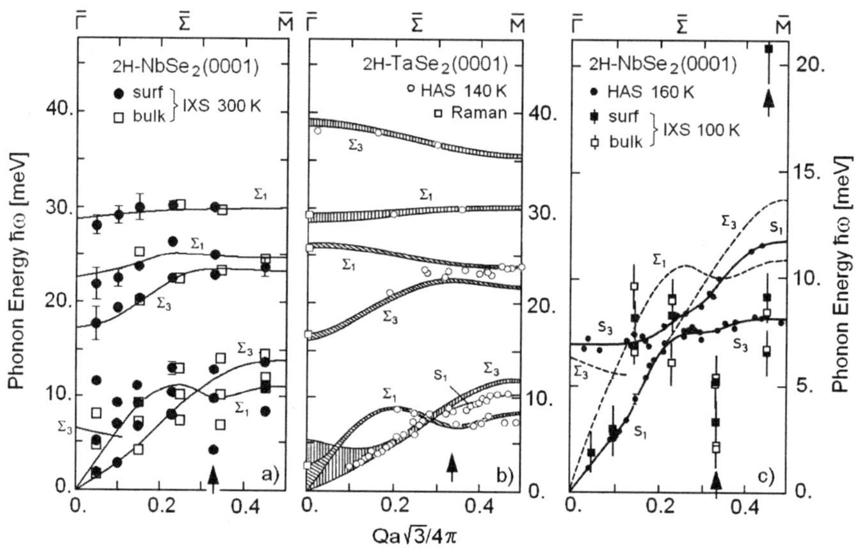


Fig. 11. (a) The surface (●) and bulk (□) phonon dispersion curves of 2H-$NbSe_2$(0001) measured with IXS at room temperature in the ΓM direction (Murphy *et al* [46,51]). The continuous curves are fits of neutron bulk data [58] for acoustic modes and eye guidelines for the optical modes. For comparison the surface phonon branches of 2H-$TaSe_2$(0001) and 2H-$NbSe_2$(0001) measured with inelastic helium atom scattering [57,56] at 140 K and 160 K, respectively, are shown in (b,c), together with the surface-projected bulk phonon bands (shadowed areas in (b)) of 2H-$TaSe_2$ as derived from a dispersive linear chain model fitted to Raman data [57]. In (c) the HAS data for 2H-$NbSe_2$(0001) are compared with neutron bulk data (broken lines) [58] and with IXS surface and bulk data at a lower temperature (100 K) [54] showing the development of the deep Kohn anomaly at about 2/3 of the zone (↑).

The first successful measurements of surface phonons with IXS have been reported by Murphy *et al* for the layered crystals 2H-$NbSe_2$(0001) [46,54] (Fig. 11), and also announced for the iso-structural 2H-$TaSe_2$(0001) [55]. These surfaces have also been investigated with inelastic helium atom scattering [56,57] so that a direct comparison is possible. Moreover a direct comparison of surface to bulk phonon dispersion curves under identical conditions, which is possible with IXS, should support the conjecture, confirmed for graphite (see above), of a negligible difference between the two sets of curves. This does not seem to be the case, however. The 2H forms of Niobium and tantalum di-selenide are known to show an incommensurate CDW phase transition at 33.5 K and 122.3 K, respectively [58]. 2H-$TaSe_2$ exhibits a lock-in transition to a commensurate phase at 90 K, whereas 2H-$NbSe_2$ remains incommensurate until it becomes superconducting at 5 K, but shows a complete softening of the acoustic longitudinal ($\Sigma_1$) mode at about 2/3 of the zone in the ΓM direction, with a second-order phase transition at 32 K [59]. On the contrary in 2H-$TaSe_2$ the frequency of this mode undergoes a substantial softening with a sharp minimum at the CDW transition without vanishing. The IXS bulk data for 2H-$NbSe_2$ at room temperature (Fig. 11(a)) show a set of acoustic and optical branches similar to the bulk branches of 2H-$TaSe_2$ (Fig. 11(b)). The optical branches of 2H-$NbSe_2$ and 2H-$TaSe_2$ fall in the same energy range and are very similar because Se atoms are mostly involved inthese modes (only the highest $\Sigma_3$ branch has not been detected in 2H-$NbSe_2$ with IXS). The IXS acoustic branches of 2H-$NbSe_2$ are instead stiffer than in 2H-$TaSe_2$ due to a pure isotope re-scaling, both showing for $\Sigma_1$ the typical Kohn anomaly at ~2/3 of the zone. It is however surprising that three acoustic branches (□) are seen instead of two, the third one showing a much softer Kohn anomaly (vertical arrow) than found with neutrons at 300 K [58].

The 2H-$NbSe_2$(0001) room temperature IXS surface data for optical phonons (Fig. 11(a),●) are practically coincident with the bulk data (□), as expected for a layered crystal and found in 2H-$TaSe_2$(0001). On the contrary the lowest surface acoustic branch shows an even deeper Kohn anomaly than in the bulk. At 100 K (Fig. 11(c)) IXS gives just two acoustic branches for the bulk as well as the surface, with deep Kohn anomalies in all of them at 2/3 of the zone. This is expected at low temperature; however, HAS data at 160 K show indeed a softening at large wavevectors of both surface branches with respect to the corresponding bulk bands, which is however quite smaller than that seen in IXS experiments. A similar softening of the $S_1$ surface branch with respect to bulk $\Sigma_3$ was observed also in HAS data for 2H-$TaSe_2$(0001) (Fig. 11(b)). Here the $S_1$ surface (Rayleigh) branch was also seen to develop a Kohn anomaly at about ½ of the zone with a minimum at 110 K [60], instead of the bulk CDW transition temperature of 122.3 K. This further anomaly is apparently not seen with IXS [55], although the effect is just within current HAS resolution and may not yet be accessible to IXS spectroscopy.

Although the first successful measurements of surface phonon dispersion curves with inelastic X-ray scattering are revealing some unexpected features not found with other more conventional methods, and wait for a confirmation and/or an explanation, it may be safely concluded that IXS has all the potential for becoming a choice tool in surface phonon spectroscopy due to the possible association with other more standard surface characterization methods currently used at synchrotron radiation facilities.

**References**


[1]) Visiting scientist.

[1] J. W. Strutt (Baron Rayleigh), Proc. London Math. Soc., **17**, 4 (1885); see also: *The Theory of Sound* (MacMillan, London 1896; Dover Publ., New York 1945).

[2] A. Gabriele, private communication (1981). Dr. Alberto Gabriele (19XX-1999), physicist, has been for many years the untiring director of the Erice Seismic Network (ESN) at the Ettore Majorana Centre for Scientific Culture, Erice, Italy. After his premature death, the ESN has been named to his memory.

[3] G.W. Farnell, in: *Physical Acoustics,* Vol. 6, Eds. W.P. Mason and R.N. Thurston (Academic Press, New York, 1970) p. 109.

[4] J. Harvey: *Sounding Out the Sun*, Physics Today, October 1995, p. 32.

[5] R. A. Broglia, Surface Sci. 500 (2002) 759.

[6] P. Hess: *Surface Acoustic Waves in Materials Science*, Physics Today, March 2002, p. 42.

[7] D. Shilo and E. Zolotoyabko, J. Phys. D: Appl. Phys. 36 (2003) A 122.

[8] A. A. Oliner (Editor): *Acoustic Surface Waves* (Springer V., Berlin, Heidelberg 1978). See particularly the articles by G. W. Farnell (p. 13), H. M. Gerald (p. 61), and E. A. Ash (p. 97).

[9] A.A. Maradudin, in: *Nonequilibrium Phonon Dynamics*, Ed. W. E. Bron (Plenum, New York 1985) p. 395.

[10] A. A. Maradudin and G. I. Stegeman, in: *Surface Phonons*, Eds. F.W. deWette and W. Kress, Springer Ser. Surf. Sci., Vol. 27 (Springer, Berlin, Heidelberg 1991) p. 5.

[11] C. K. Campbell, *Surface Acoustic Wave Devices for Mobile and Wireless Communications* (Academic, New York, 1998).

[12] K.-Y. Hashimoto, *Surface acoustic wave devices in telecommunications modeling and simulation* (Springer, Berlin, 2000).

[13] C. C. W. Ruppel and T. A. Fjeldly (Editors), *Advances in Surface Acoustic Wave Technology, Systems and Applications,* Vol. 1 (World Sci., Singapore 2000); Vol. 2 (World Sci., Singapore 2001).

[14] C. C. W. Ruppel, L. Reindl, R. Weigel, IEEE Microwave Magazine, 3 (2002) 65.

[15] D. Morgan, *Surface Acoustic Wave Filters,* 2nd Ed. (Academic Press, 2007)

[16] L. Eichinger, S. Sischka, G. Olbrich, R. Weigel, Proc. IEEE Ultrasonics Symposium (IEEE, Piscataway, NJ 2000) p. 29.

[17] L. Reindl, C.C.W. Ruppel, S. Berek, U. Knauer, M.Vossiek, P. Heide, and L. Oreans, IEEE Transactions on Microwave Theory and Techniques, 49 (2001) 787.

[18] S. Scheiblhofer, S. Schuster, and A. Stelzer, IEEE Transactions on Microwave Theory and Techniques, 54 (2006) 1477.

[19] See, e.g., the papers in the *Special Issue on Acoustic Wave Sensors and Applications,* IEEE Transact. Ultrasonics, Ferroelectrics and Frequency Control, 51 (2004) 1365-1469.

[20] M. M. de Lima Jr., M. Beck, R. Hey, and P. V. Santos, Appl. Phys. Letters, 89 (2006) 121104.

[21] V. Laude, M. Wilm, S. Benchabane, and A. Khelif, Phys. Rev. E 71 (2005) 036607

[22] S. Benchabane, A. Khelif, J.-Y. Rauch, L. Robert, and V. Laude, Phys. Rev. E 73 (2006) 065601.

[23] O. B. Wright and K. Kawashima: Phys. Rev. Lett. 69 (1992) 1668.

[24] Y. Sugawara, O. B. Wright, O. Matsuda, M. Takigahira, Y. Tanaka, S. Tamura, and V. E. Gusev, Phys. Rev. Lett. 88 (2002) 185504.

[25] D. M. Profunser, O. B. Wright, and O. Matsuda, Phys. Rev. Lett. 97 (2006) 055502.

[26] C. Rocke, S. Zimmermann, A. Wixworth, J. P. Kotthaus, G. Bohm and G. Weinman: Phys. Rev. Lett. 78 (1997) 4009.

[27] V. I. Talyanskii, J. M. Shilton, M. Pepper, C. G. Smith, C. J. B. Ford, E. H. Linfield, D. A. Ritchie and G. A. C. Jones: Phys. Rev. B 56 (1997) 15180.

[28] M. Kataoka, C. H. W. Barnes, H. E. Beere, D. A. Ritchie and M. Pepper, *Phys. Rev. B* 74 (2006) 085302.

[29] H. Totland, Ø. L. Bø and Y. M. Galperin: Phys. Rev. B 56 (1997) 15299.

[30] G. D. Mahan: *Many-Particle Physics* (Plenum Press, New York 1981) Sec. 6.2.

[31] N.-H. Ge, C. M. Wong, R. L. Lingle, Jr., J. D. McNeill, K. J. Gaffney and C. B. Harris, Science 279 (1998) 202.

[32] U. Höfer: Science 279 (1998) 190.

[33] A. B. Hutchinson, V. I. Talyanskii, M. Pepper, G. Gumbs, G. R. Aizin, D. A. Ritchie, and E. H. Linfield, Phys. Rev. B 62 (2000) 6948.

[34] M. Kataoka, R. J. Schneble, A. L. Thorn, C. H. W. Barnes, C. J. B. Ford, D. Anderson, G. A. C. Jones, I. Farrer, D. A. Ritchie, and M. Pepper, Phys. Rev. Lett. 98 (2007) 046801.

[35] J. Ebbecke, N. E. Fletcher, T. B. M. Janssen, F. J. Ahlers, M. Pepper, H. E. Beere, and D. A. Ritchie, Appl. Phys. Lett. 84 (2004) 4319.

[36] R. Rodriquez, D. K. L. Oi, M. Kataoka, C. H. W. Barnes, T. Ohshima, and A. K. Ekert, Phys. Rev. B, 72, (2005) 085329.

[37] M. Cecchini, G. De Simoni, V. Piazza, F. Beltram, H. E. Beere, and D. A. Ritchie, Appl. Phys. Lett., 85 (2004) 3020.

[38] M. Cecchini, V. Piazza, F. Beltram, D. G. Gevaux, M. B. Ward, A. J. Shields, H. E. Beere, and D. A. Ritchie, Appl. Phys. Lett. 86 (2005) 241107.

[39] C. H. W. Barnes, J. M. Shilton, and A. M. Robinson, Phys. Rev. B, 62 (2000) 8410.

[40] C. H. W. Barnes, Philos. T. Roy. Soc. A, 361 (2003) 1487–1492.

[41] S. Furuta, C. H. W. Barnes, and C. J. L. Doran, Phys. Rev. B, 70 (2004) 205320.

[42] K. Ishioka, M. Hase, M. Katajima, and H. Petek, Appl. Phys. Lett. 89, 231916 (2006).

[43] K. Ishioka, M. Hase, M. Kitajima, L. Wirtz, A. Rubio and H. Petek, Phys. Rev. Lett. (2007).

[44] Y. Sugawara, O. B. Wright, and O. Matsuda, Appl. Phys. Lett. 83, 1340 (2003).

[45] G. Alexandrowicz, A. P. Jardine, H. Hedgeland, W. Allison and J. Ellis, Phys. Rev. Lett. 97 156103 (2006).

[46] B. M. Murphy, H. Requardt, J. Stettner, M. Müller, M. Krisch, and W. Press, Phys. Rev. Lett. 95, 256104 (2005)

[47] J. R. Dutcher, S. Lee, B. Hillebrands, G. J. McLaughlin, B. G. Nickel and G. I. Stegeman, Phys. Rev. Lett. 68, 2464 (1992).

[48] I. Giovannini, F. Nizzoli and A.M. Marvin, J. Electr. Spectr. Rel. RW, L Phenom. 64/65 (1993) 783.

[49] G. Benedek, F. Traeger, and J. P. Toennies, Phys. Rev. Lett. 94, 086103 (2005).

[50] A. Glebov, W. Silvestri, J. P.Toennies, G. Benedek and J. G. Skofronick, Phys. Rev. B 54, 17866 (1996).

[51] J. Maultzsch, S. Reich, C. Thomsen, H. Requardt. and P. Ordejòn, Phys. Rev. Lett. 92, 075501 (2004).

[52] J. Maultzsch, S. Reich, and C. Thomsen, Phys. Rev. B 70, 155403 (2004).

[53] G. Benedek and G. Onida, Phys. Rev. B 47, 16471 (1993-II).

[54] B. M. Murphy, M. Müller, J. Stettner, H. Requardt, J. Serrano, M. Krisch, and W. Press, J. Phys. Cond. Matter, to appear (2008).

[55] B. M. Murphy, private communication; see also her Thesis (University of Kiel, 2003).

[56] G. Benedek, L. Miglio and G. Seriani, in *Helium Atom Scattering from Surfaces*, ed. by E. Hulpke (Springer V., Heidelberg 1992) p. 207] .

[57] G. Benedek, L. Miglio, G. Brusdeylins, C. Heimlich, J.G. Skofronick and J.P. Toennies, Europhys. Lett. 5 (1988) 253.

[58] D. E. Moncton, J. D. Axe, F. J. DiSalvo, Phys. Rev. Lett. 34, 734 (1975).

[59] C. Ayache, R. Currat, and P. Molinié, Physica 180 (1992) 333.

[60] G. Benedek, G. Brusdeylins, G. Heimlich, L. Miglio, J.G. Skofronick, J.P. Toennies and R. Vollmer, Phys. Rev. Lett. 60 (1988) 1037.